A House built on Sand

- a political economy of Saudi Arabia

Helen Lackner

A House Built on Sand
a political economy of
Saudi Arabia

by Helen Lackner

Ithaca Press

DS
244.52
.L3
c.2

© 1978 Helen Lackner

First published in 1978
by Ithaca Press
13 Southwark Street London SE1

ISBN 0 903729 27 X Cloth
ISBN 0 903729 28 8 Paperback

Photoset by FI Litho Ltd London N1

Printed in England
by Anchor Press Ltd and bound
by Wm Brendon & Sons Ltd,
both of Triptree Essex

Contents
Maps
Family Tree of King Abdel Aziz
Preface   i
Publisher's Note   iv

I Arabia up to 1900	1
II The Formation of Saudi Arabia	14
III Oil: from Discovery to International Politics	32

IV The Contemporary State
   1 Politics & the Ruling Family
   2 State Structures
	57

V Political Opposition	89
VI Saudi Arabian Foreign Policy	110
VII The Economy & the Development Plans	137
VIII Saudi Arabian Society in Transition	172
Conclusion	213
Further Reading	218
Index	219

Tables

Currency equivalents	iv
Crude Oil Productuion	44
Oil Revenues & Reserves	45
Number of Boys & Girls in State Schools	79
Foreign Aid committed by Saudi Arabia	130
Saudi/US balance of trade	134
Government Income & Expenditure 1966-69	140
Summary of 5-year Plan allocations	141
Employment in Manufacturing Industry	144
The Plan's Estimated Recurrent & Project Costs	152
Main Development Programmes	154
Estimated Manpower by Occupational Group	164
Agricultural production by crop	183
Length of service of Saudi employees in Aramco	189

To the memory of my father,
Joachim Lackner,
who struggled against fascism and oppression
throughout his life.

CYPRUS

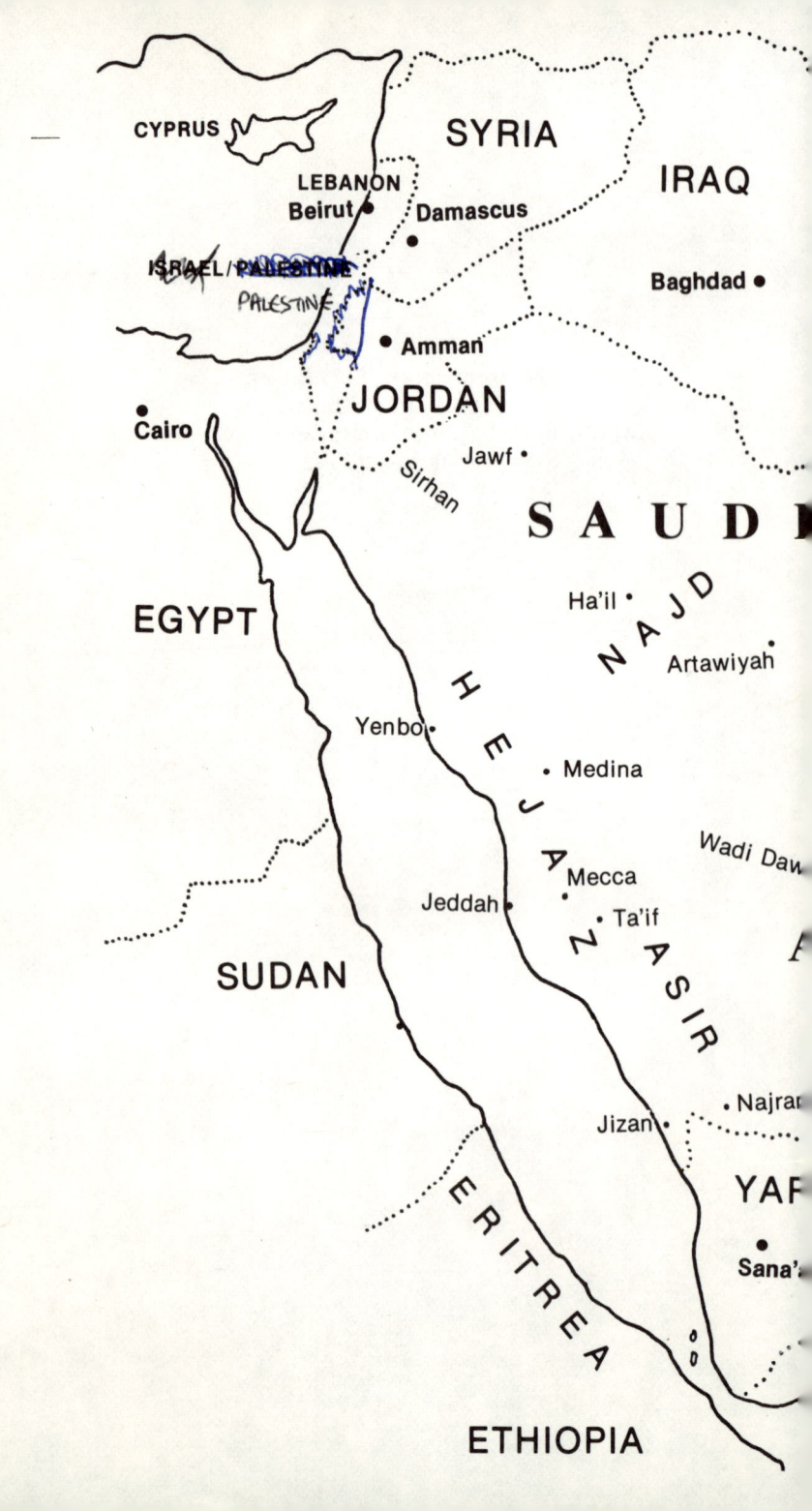

SYRIA

IRAQ

LEBANON
Beirut · · Damascus

ISRAEL/PALESTINE
PALESTINE

Baghdad ·

· Amman

JORDAN

· Cairo

Jawf ·

Sirhan

SAUDI

EGYPT

Ha'il ·

NAJD

· Artawiyah

Yenbo ·

H
E
J
A
Z

· Medina

Wadi Daw

Jeddah ·

· Mecca

· Ta'if

SUDAN

A
S
I
R

· Jizan

· Najra

ERITREA

YA

· Sana'

ETHIOPIA

IRAN

KUWAIT

Jubail
· Qatif
Dammam · BAHRAIN
· Dhahran
al Hasa Hofuf QATAR
YADH   Abu Dhabi · UNITED ARAB EMIRATES
AL KHARJ   UNITED ARAB EMIRATES   Muscat
AJ

OMAN

A B I A

DHOFAR

HADRAMAUT
PDRY

| 0 | 500 | 1000Km |

Map of Arabian Tribes

- ARUWALAH
- ANAZAH
- AL-'AMARAT
- AL-MUNTAFIQ
- BANI SAKHR
- *Wadi as-Sirhan*
- BANI 'ATIYAH
- AS-SULABAH
- ADH-DHAFIR
- AL-HUWAYTAT
- 'ANAZAH
- SHAMMAR
- *Wadi al-Batin*
- 'ANAZAH
- SHAMMAR
- BALI
- HUTAYM
- HARB
- *Wadi ar-Rumah*
- JUHAYNAH
- 'UTAYBAH
- SUBAY'
- Riyadh
- al-Qa'iyah
- Medina
- AD-DAWASIR
- AS-S
- Layla
- *Wadi Fatimah*
- AL-BUQUM
- Jeddah
- Mecca
- Ta'if
- SUBAY'
- GHAMID
- AD-DAWAS
- ZAHRAN
- *Wadi ad-Dawasir*
- HUDHAYL
- QAHTAN
- SHAHRAN
- 'ASIR
- YAM
- Najran
- AL-AMALISAH
- WAYILAH
- DA
- San'a
- AL-'AWAL
- YAFI'
- Aden

Tribal Map of the Arabian Peninsula

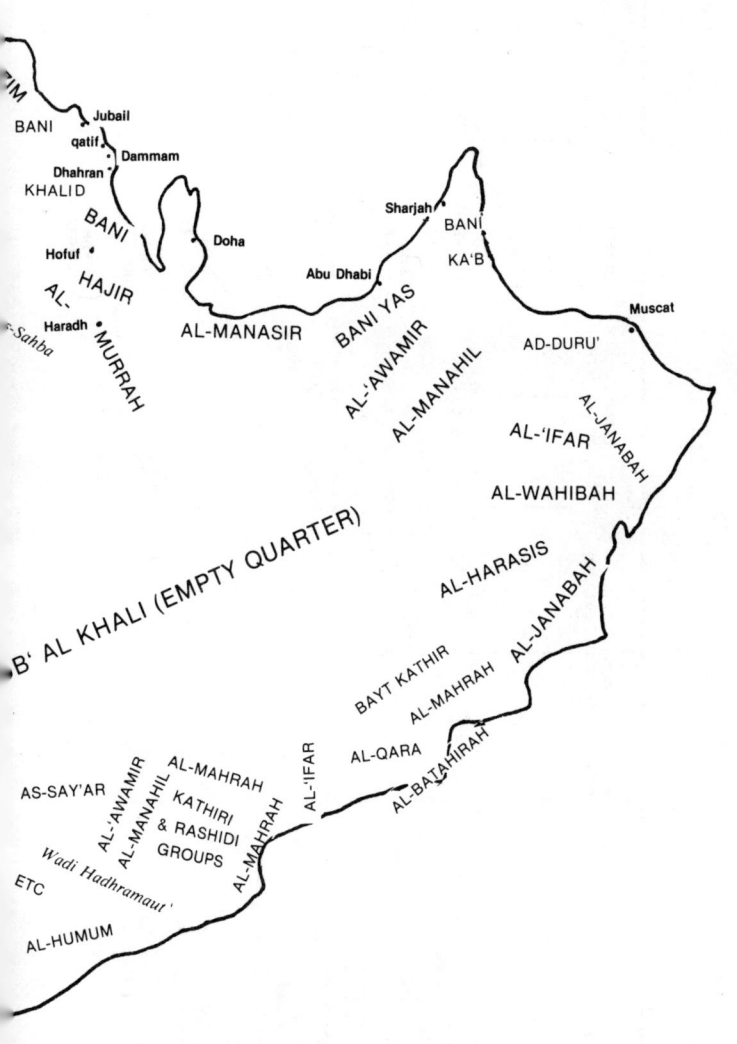

Adapted from ARAMCO Handbook, Dhahran 1968.

THE FAMILY OF KING ABDEL AZIZ*
(Mothers' names in parentheses)

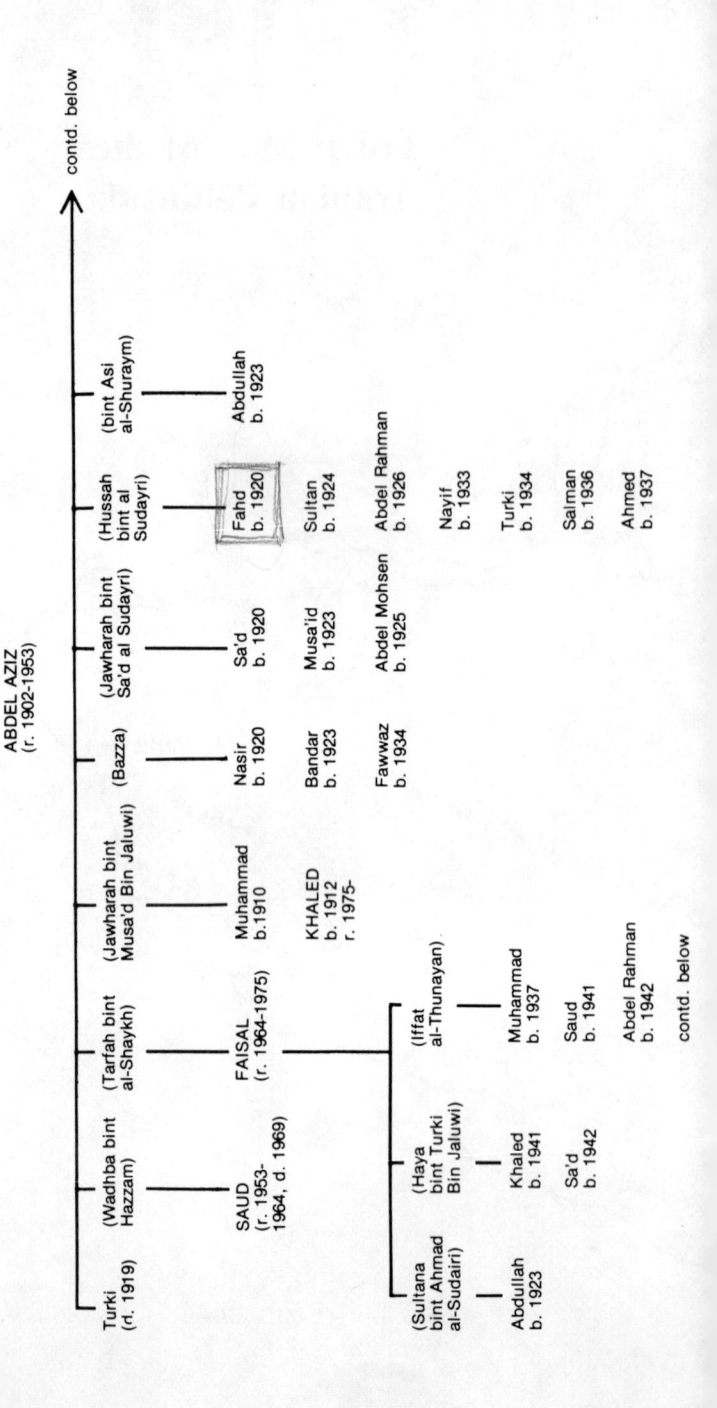

ABDEL AZIZ
(r. 1902-1953)

- Turki (d. 1919)
- (Wadhba bint Hazzam)
  - SAUD (r. 1953-1964, d. 1969)
    - (Sultana bint Ahmad al-Sudairi)
      - Abdullah b. 1923
    - (Haya bint Turki Bin Jaluwi)
      - Khaled b. 1941
      - Sa'd b. 1942
    - (Iffat al-Thunayan)
      - Muhammad b. 1937
      - Saud b. 1941
      - Abdel Rahman b. 1942
      - contd. below
- (Tarfah bint al-Shaykh)
  - FAISAL (r. 1964-1975)
- (Jawharah bint Musa'd Bin Jaluwi)
  - Muhammad b. 1910
  - KHALED b. 1912, r. 1975-
- (Bazza)
  - Nasir b. 1920
  - Bandar b. 1923
  - Fawwaz b. 1934
- (Jawharah bint Sa'd al Sudayri)
  - Sa'd b. 1920
  - Musa'id b. 1923
  - Abdel Mohsen b. 1925
- (Hussah bint al Sudayri)
  - **Fahd b. 1920**
  - Sultan b. 1924
  - Abdel Rahman b. 1926
  - Nayif b. 1933
  - Turki b. 1934
  - Salman b. 1936
  - Ahmed b. 1937
- (bint Asi al-Shuraym)
  - Abdullah b. 1923

contd. below

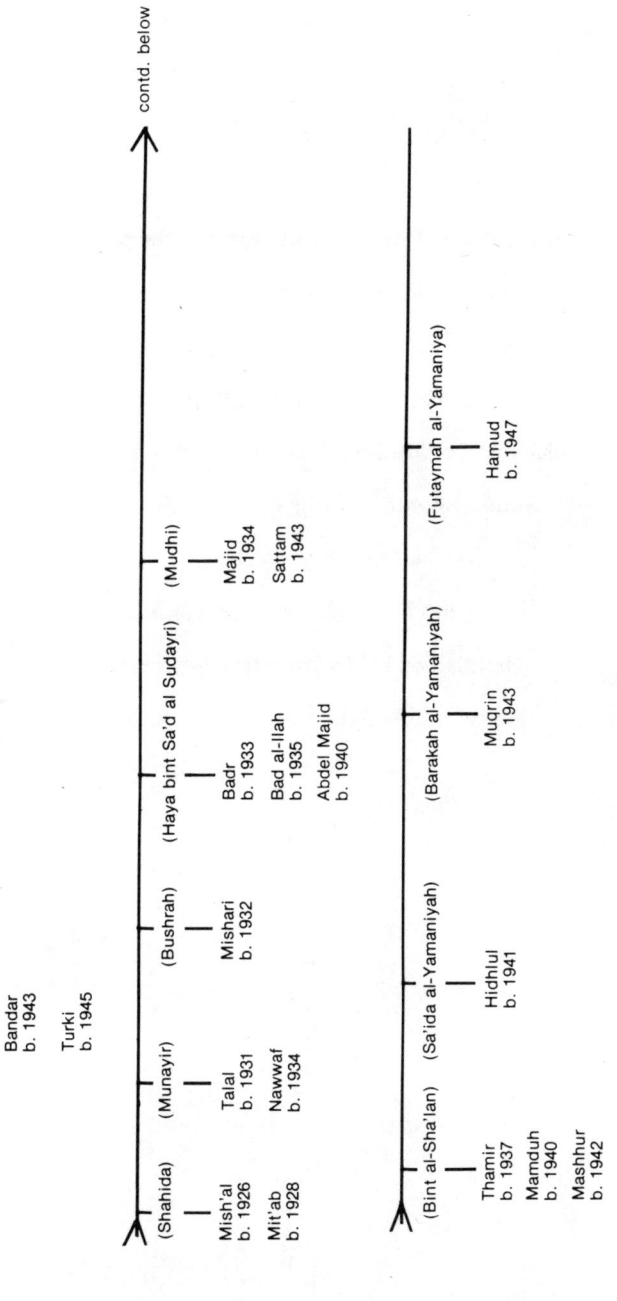

Sons of Iffat & Faisal
continued

↑ contd. below

(Shahida)   (Munayir)   (Bushrah)   (Haya bint Sa'd al Sudayri)   (Mudhi)

Mish'al   Talal   Mishari   Badr   Majid
b. 1926   b. 1931   b. 1932   b. 1933   b. 1934

Mit'ab   Nawwaf   Bad al-Ilah   Sattam
b. 1928   b. 1934   b. 1935   b. 1943

                                           Abdel Majid
                                           b. 1940

Bandar
b. 1943

Turki
b. 1945

↑

(Bint al-Sha'lan)   (Sa'ida al-Yamaniyah)   (Barakah al-Yamaniyah)   (Futaymah al-Yamaniya)

Thamir   Hidhlul   Muqrin   Hamud
b. 1937   b. 1941   b. 1943   b. 1947

Mamduh
b. 1940

Mashhur
b. 1942

*With the exception of the first three, the chart includes only surviving sons.

Adapted from David E. Long, **Saudi Arabia**, Washington Papers Vol IV No 39 1976, Beverly Hills & London, Sage Publications.

The Land Where Power Springs From A Barrel Of Oil

Saudi Police Captain Fined For Theft

Spending $142,000 Million Is Not So Easy

King Warns Women Over Bare Legs

Khaled To Buy Jumbo Jet For His Own Use

Saudis Should Pay For Pentagon Survey

Saudis Stoned To Death

US Bribes To Saudi Generals Admitted

Saudis Top US In Monetary Reserves

'Sexist' Saudis May Not See The Queen'

PREFACE

Saudi Arabia makes good headlines. These are just a few examples, by no means the most sensational, of recent newspaper headlines. Until recently an obscure Arab monarchy, the OPEC oil price rises of 1973 and the oil embargo of that year transformed the situation overnight. Saudi Arabia became a country of world significance. In the Middle East, its financial power has given Saudi Arabia dominance over the confrontation states in the struggle for Palestine. On the world scene, Saudi Arabia is in a position to assist Third World countries in the struggle against imperialism, but despite a number of statements in support of improved terms of trade and credit facilities, Saudi Arabia has done little. In the West, Saudi Arabia's accumulated financial reserves have given it second place after West Germany in the world's league of foreign reserves.

Saudi Arabia's ambitious Development Plan, announced in 1975, has given it further opportunities to support Western economies in their moment of crisis, as it involves spending $142,000 million, excluding all military contracts. Most of this is expected to go to Western firms and thus 'recycle' petrodollars, a euphemism invented to give an aura of respectability to capitalist efforts aimed at ensuring that any money spent on oil imports is repatriated as fast as possible in sales of questionable value.

The most public appearance of Saudi Arabia in the Western world has been in the carefully nurtured 'corrupt big spender' image touted in the media. Interest in the way of life of oil sheikhs is combined with envy at their *nouveau riche* spending habits in the West, their purchases of stately homes and their depraved habits such as gambling and frequenting night clubs. There is little or no discussion of Saudi Arabia's internal social structure and dynamics.

The aim of this book is to contribute towards a better understanding of Saudi Arabia's internal situation and to correct some of the misconceptions prevalent about the country. Unlike other contributions on Saudi Arabia, this one is primarily concerned with the fate and interests of the indigenous population of the country, and more particularly with its less wealthy members. This book is not designed to assist Western businessmen in making greater profits from Saudi Arabia's oil wealth. Nor will it give Saudi Arabians a blue print for their development. Instead it tries to present the problems which are inherent to the political, economic and social policies which currently exist, and to provide the elements of a discussion on possible alternatives.

Among developing countries, Saudi Arabia is unique; no analysis of it can fit into any pre-conceived model of development as the country's specificity far overrides any features it may share with other developing nations. Like most other Third World states, its borders were recently defined, but unlike many it was never colonised: therefore its traditional mode of production lasted well into the 20th. Century, while that of most underdeveloped countries was destroyed in the 19th. Similarly, the population's achievement of 'nationhood' is still somewhat questionable, though far less so than it was 20 years ago.

The biggest difference between Saudi Arabia and most Third World 'developing' countries rests on two factors: unlike other Third World countries, Saudi Arabia is rich: its oil revenues are astronomical and it therefore need not be dependent on imperialism for investment capital, though it remains dependent for the importation of technology and labour. Saudi Arabia has a very small population: officially claimed to be 6 million or above, the population is more reliably estimated at between 3.5 and 4 million indigenous inhabitants. This has considerable implications for development as it precludes the initiation of productive activities on the basis of labour intensive projects. The state's repressive social policies exacerbate the shortage of labour power by excluding women from the labour force, leaving just over 1 million people available for work.

Apart from oil and other minerals, Saudi Arabia's resources are limited and its infrastructure is still very inadequate. Its agricultural potential is small and its industry almost non-existent. It has unlimited investment capital but a very small population and few resources which could be autonomously developed without considerable imported inputs. Its development problems, therefore, cannot be compared to those of other countries. By contrast with many developing states which have to rely on their human potential for development, Saudi Arabia is in a position to finance any development options it may select but has to find policies suited to its unique conditions, which open fundamental questions about the nature of development. Is industrialisation necessary to create a developed economy? If so, what kind of industrialisation could be viable in Saudi Arabia? Could Saudi Arabia become a rentier state, satisfying its population's needs from the state's income from foreign investment? This book is concerned with such questions from the point of view of the welfare of the Saudi Arabian population, not that of Western capitalism.

The other concerns of this book are the absence of political life in the country and the origins of the present political structure, as well as its long-term viability on the one hand and the transformation of Saudi Arabian society on the other. The uprooting of the traditional social structure is only a fairly recent phenomenon and the social formation which will replace it only embryonic, but certain trends are already visible.

Writing about Saudi Arabia is difficult. The uncertainty over the population figures render much other statistical information questionable, and the state's organisation of available information is sometimes fanciful. In my work I have greatly benefitted from discussions with, and information from many people. I want first and foremost to thank all the people in the Middle East who have helped me but who I am sure would prefer to remain anonymous. In London I want to thank the following for information, criticisms, suggestions, and for reading the manuscript at various stages of preparation: Nigel Disney, John Gittings, Fred Halliday, Pamela Ann Smith and Dave P. Taylor. Dorothy Mbeki helped me edit the final manuscript and the Transnational Institute (Amsterdam) gave me financial support in the early stages of my research. Ken Whittingham financed the writing and contributed in innumerable ways throughout. I am grateful to them all and to Panda Cat, for its furry presence. I hope that the analysis which is presented here will contribute to a more sympathetic understanding of Saudi Arabia, its fascinating and exciting problems. I take full responsibility for any factual errors and the interpretation presented here.

London, October 1977

PUBLISHER'S NOTE

Saudi Arabian currency is the Saudi Riyal (SR) which until January 1960 was tied to the US dollar ($) at the rate:

from Jan 60 to Aug 71	SR4.50/$
from Dec 71 to Feb 73	SR4.145/$
from Feb 73 to Aug 73	SR3.730/$
from Aug 73 to Mar 75	SR3.55/$
from Aug 75	SR3.53/$

Since March 75 the Riyal was tied to the IMF special Drawing Right unit (SDR) — itself based on a basket of 16 currencies — at the rate:
SR4.28255/SDR with fluctuations to allow a fixed dollar rate to be held for extended periods. From November 67 to June 72 the sterling rate was SR10.80/£. (From figures given in *Europa Handbooks: Middle East & N. Africa* 1977.)

Values are quoted in sterling only for those British political subventions pre-World War II. International contracts and trade are quoted in US $ and internal budgets etc in Saudi Riyals.

While consistency has been sought throughout the text for names transliterated from Arabic, the reader should note that there is no universally accepted system of roman spelling. For example, Amarate is the same as Emirate; Jiddah equals Jeddah; Fahad equals Fahd.

Areas are quoted in Dunum which vary in size throughout the Middle East. In Saudi Arabia it may be taken equivalent to one tenth of a hectare.

CHAPTER I
ARABIA UP TO 1900

Saudi Arabia occupies about 80 per cent of the Arabian peninsula, bordered by Jordan, Iraq and Kuwait to the north, the Red Sea and the Yemen Arab Republic to the west, the People's Democratic Republic of Yemen and Oman to the south, the United Arab Emirates and the Gulf Coast to the east. As they are largely in the desert, its borders have not all been precisely determined. The country covers about 1,400,000 square kilometres, less than 1 per cent of which is cultivated: the rest is mainly desert with some areas suitable for occasional camel grazing. There are no permanent rivers, only *wadis* which are filled with torrents of floodwater during the occasional rains, but are otherwise dry. Wells and occasional springs are the other main sources of water: larger springs sometimes develop into oases where people can settle permanently and live off agriculture.

From the Red Sea coast, the land rises sharply into a plateau which then slopes gently towards the east, leaving the steepest areas near this coast. This escarpment forms two mountain ranges: the first in the Hejaz, separated from the second in Asir by a gap located in the Mecca region. East of these mountains is the great plateau of Najd, which includes grazing land in the northern parts, and the Nafud and Dahna deserts; in the rest of Najd, grazing land is scattered with occasional oases, mainly along *wadis* and mountain ranges. The Tuwaiq range separates Najd from the Rub' al Khali, the largest and least hospitable of the Arabian deserts.

The eastern region of the peninsula is composed of the flat Gulf coastal plain which is largely barren and of the al Hasa region formed of two major oases near the town of Hofuf which are the centres of date cultivation, with 2 million date palms and about 30,000 acres of irrigated arable land.

Climatically the peninsula is particularly inhospitable. The Hejaz is extremely hot and humid while Asir is the only region which receives enough of the monsoon rains to make agriculture without irrigation possible. The rest of the country is hot by day and cold at nights. There is very little rain, under 10cm. a year, which falls in winter if at all.

Traditional Political Economy of the Region

In the interior of the peninsula, nomadic camel-herding and peasant agriculture in the oases persisted from pre-Islamic times into the 20th. Century. They were means of survival well adapted to the scarcity of resources and while there were trade relations between peasant and nomad these did not provide the basis for transforming the mode of production in the interior. Market towns developed where the trade routes met and these, together with the coastal strips provided links between the peninsula and the outside world. The surplus which accrued to the urban merchants through trade and, with the spread of Islam, from the pilgrimage, was not however used to develop the productive forces of the peninsula — there were few around. Urban populations, nomads, and the oases people were affected by changes in the forms and amount of trade and the revenue coming in with pilgrims, but it is only in the towns that the degree of dependency on trade relations and pilgrim taxation was crucial. Before the discovery of oil there was more possibility of a regression in the level of development in the peninsula, whether through prolonged drought, decline of trade or number of pilgrims, than of transformation to a higher level.

The peninsula in these conditions was not a unified society; while the tribal structure and, later, Islam divided oasis, desert and urban communities; the degree of isolation from each other and the mode of production made the possibility of a unified social and political structure meaningless. This could arise only by the destruction of the old forms and the integration of the desert and oasis people into a single economy transcending simple trade relations. It is only in the 20th. Century, under political pressures from the outside world that the creation of a common political structure for the whole region was placed on the agenda, and only with the discovery of oil and its economic effects throughout the peninsula that the centralisation of state power became both a possibility and a necessity.

In the centuries before the 20th. Century there was little change in the four main sectors of economic life: agriculture in Asir and the oases, fishing for food and pearls on the Gulf Coast, camel herding in the desert and trading and handicrafts in the towns.

Agriculture

A cursory glance at the map reveals the degree of isolation of these peasant communities. The main areas of agricultural settlement were the Tihama plain in the west and the hills of Asir which were integrated economically and culturally with the Yemen. Since this region received monsoon rainfall, crops could be cultivated without irrigation and dams built to make irrigation possible in the drier parts. The hills were terraced to make maximum use of the fertile volcanic soil and the rainfall. The crops were varied, and included vegetables, grain and fruit. Unique to this region were coffee, indigo and *qat* (a bush whose leaves, when chewed, produce a mild narcotic effect), all of which were important in the economy of Yemen.

In the desert regions, agriculture was practised primarily in the oases of Najd and in al Hasa near the Gulf coast. Dates were the basis of oasis life, the staple of human diet, eaten raw, cooked, compressed, fresh, dried or in syrup. The tougher varieties were used as animal fodder, the stones to feed camels, the fronds for mats, rope making, housing and household implements, and the trunks for house building and fuel. Dates were the only product grown in sufficient quantities to be traded outside the oasis community. A few vegetables and grain were also grown and some goats and sheep kept in the oases areas.

Ownership of trees, particularly date palms, was of some importance in the oases: sometimes they belonged to individual peasant families and sometimes to nomad families who came to harvest the dates. Access to the water supply was a major source of conflict in the oases as water was the most scarce and essential resource. The community appointed a supervisor to ensure that each family used water only for the amount of time they were entitled to. Since the richer families could afford more water, this was a way of cementing their relative position.

Land was mainly owned by families who worked on it themselves. It was rare to find large landowners who employed peasants on a permanent basis. Small family plots tended to be divided to a point below economic viability by the inheritance system: one solution to this problem was to turn the ownership of the land over to religious *waqf* or trusts under terms which gave its use and product to the original family and avoided the division of the plot at every generation. Inequalities in the ownership of land and date palms, as well as of access to water, created inequalities between the families in the peasant community, with the economically dominant family tending to dominate the council of elders who decided the common affairs of the community. However the low level of economic resources limited the degree of inequality which could develop, leaving rich and poor on familar terms unlike modern society.

The nomads

The camel was for the nomad what the date was for the peasant. The camel provides milk and its by-products as food, and occasionally meat. Camel hair is used for clothing and tenting, its dung for fuel and its urine as shampoo, and its skin for containers and sandals. Most importantly the camel is a safe and reliable means of desert transport: it can survive for up to three weeks without water. For the *bedu* the camel provided both the means of subsistence and the means of trading.

Each year the nomads' travelling cycle followed the same pattern: winter was spent travelling through the desert regions in search of the short lived grazing within their traditional territory, and summers were spent near a permanent water hole, or an oasis where they had relations with the villagers or perhaps owned palms.

Individual property rights in land were meaningless for the nomadic communities. Tribes and sections of a tribe had identified and recognised areas which were considered to be theirs and which included wells and water holes. The delimitation of these areas was somewhat flexible and changed according to grazing conditions and agreements between tribes, i.e. a tribe whose area was particularly barren could make an arrangement whereby a neighbouring allied tribe would allow them to graze their camels in their region, which had received more rainfall. Frequently there was no agreement, and then the use of another group's land led to inter-tribal conflict. Such fighting also took the form of camel stealing as a means of survival, or as bridewealth or for sale to neighbouring regions. Raids and counter-raids for camels were part of the mythic as well as basic economic and political life of the desert, and were an expression of the continuous struggle for existence.

The life of the interior of the peninsula was dominated by the *bedu*. They exchanged their camel products for dates in the oases, and exchanged camel products and dates not consumed by themselves in the towns for imported grain and weapons. These weapons and the mobility provided by the camel enabled the nomads to dominate the settled people (*hadhra*) whom they despised and did not hesitate to raid when they thought they could get away with it. The nomads formed the link for all trade and communications between the agricultural oases and the market towns. Their trading activities went beyond mere exchange of basic necessities with the villagers, or between villages and market towns: they also operated the caravan routes transporting goods between south Arabia and the Fertile Crescent, thus forming a link between the Far East, India and East Africa in one direction and the Mediterranean in the other. Until the arrival of the steamship the caravans were a vital link between Asia and Europe, transporting silks, spices and other luxuries for the rulers of Europe.

Apart from the fully nomadic groups which travelled through the most forbidding parts of the desert, there were also semi-nomads, who lived on the fringes of the desert where it was possible to find enough fodder for sheep and goats without having to travel too far for water. These *bedu* remained in closer contact with villages and towns to exchange animal products for agricultural produce and handicrafts.

The towns of Hejaz

Since antiquity the Red Sea port of Jeddah and the major towns of Mecca and Medina were trading centres where the intercontinental trade routes met. Apart from the transit of products to Europe, rice and weapons for internal consumption in the peninsula were exchanged for local dates and camel products. The fishing villages along the coast provided some food for the towns, but most supplies came from oases within easy reach.

The rise of Islam increased the importance of these towns — they were not merely market towns, but religious centres and the old trade routes also became pilgrim trails. The revenue obtained from pilgrims was of major importance to these towns, since all services had to be provided for the visitors. The pilgrimage also encouraged the development of slave markets in these towns, as some poor pilgrims who ended up destitute were then sold as slaves both to rich merchants and nomads as domestics and to the rare landowners as slaves for their farming land. Later, the pilgrimage provided a good cover for slave traders.

The Gulf Coast

Along the Gulf coast a number of small villages lived off the sea, both from fishing and pearl diving. Although the Gulf has great fish wealth, it seems that this was not exploited to its full extent and that pearls were the main resource obtained from the Gulf's waters. Pearl diving dominated the economic structure of the villages where the owners of diving *dhows* were in a powerful position. They gathered all the pearls collected in the summer's catch and sold them to the richer, larger pearl traders who either visited the villages or traded in the main centres of Bahrain and Bombay. The boat owners also had a considerable hold over the ordinary divers most of whom were totally dependent on them and, in winter, had to borrow money from them to survive, thus rapidly becoming a form of indentured labour. Once a diver borrowed from an owner, he was usually unable to repay the sum plus the interest till the end of the summer, after returning from his first year of work for his creditor. As the income was usually small the diver would have to borrow more in the following winter, thus falling into the vicious circle of indebtedness. While on the boat, all divers went into the water for up to two minutes at a time, and dived at least 10 times a day, with no more protection than a nose grip. All the pearls were gathered together and the owner would sell them and give an equal share to the divers, taking a double share for himself. The hard life of pearl divers is graphically described in the film *The Cruel Sea* by the Kuwaiti director Khaled es Seddik.

Politically the peninsula was divided into tribes and sub sections of tribes: the nomads travelled in a single lineage group (claiming a common ancestor), the largest viable travelling unit, but in the towns and the villages more than one lineage made up the community. A hierarchy of tribes and tribal sections was based on kinship relations to the eponymous ancestor of the tribe; but in many cases kinship connections were fictitious rather than real as foreign groups were integrated into the lineage over a period of time or as a result of defeat in fighting or some other disaster. The working political unit was the lineage, organised around its sheikh, usually an elderly sage chosen by the elders of the different families composing the lineage group. Although commonly described as egalitarian, insofar as positions could be acquired by merit and a leader could be deposed by the community in certain circumstances, it is clear that the leadership was virtually hereditary, being restricted to the most influential and powerful members of the most important family.

One of the duties of the sheikh was to hold a regular *majlis* — where any member of the community could raise matters of concern. If a dispute occurred between members of the community, the sheikh, having heard both parties at the *majlis*, would settle the issue in consultation with the other elders. The consensus of the elders was essential to the cohesion of the community. The tradition of open court at the *majlis* persists to this day even with the Saudi royal family.

Both in settled communities and among the *bedu* knowledge of kinship was of vital importance to political alliance. In the case of the nomads there was a complex set of alliances between tribes in case of warfare or need, some of these alliances were long term, usually between closely related sub-sections of a tribe, but most of them were on a short term basis relating to the right of way or use of grazing land and water, as well as to joint military campaigns.

Although some tribes were occasionally able to establish a temporary domination over a large area, this was never long lasting nor did it form the basis for an attempt to extend power across the peninsula. Whereas the nomadic sections of a tribe dominated the oases villages, the market towns were able to defend themselves and could confront the nomads on more equal terms. However these relationships were gradually weakened in the towns as the markets became dominated by the wealth of the merchants.

Islam

It was in this context that Islam developed in one of the principal market towns. Mecca was the town in which the Quraish tribe was based and had become a major trading centre as well as a religious centre. The Ka'aba sanctuary area had been enlarged before Muhammad's time thus affording protection to new residents. By the time of Muhammad's birth in 571, his tribe, Quraish, had been dominant in Mecca for a century, having given up nomadic life for the benefits and profits of urban trade.

In the Sixth Century the Hejaz towns grew through the increased caravan trade which moved there as a result of the disintegration of the Byzantine and Sassanian empires to the north, and the collapse of the Yemeni empire to the south. Although the nomads found it easy to dominate settled but isolated people in agricultural oases, they were no match for the towns: here they fell under the dominance of the merchants' control of trade, particularly as the merchants reduced their dependence on nomads by organising caravans to develop their business. As the towns grew, so did social and economic inequalities among the urban population.

Muhammad preached a monotheistic and egalitarian religion which appealed to the discontented and impoverished. His message spread from its city base where the Friday prayers and sermons could take place and large numbers of people could be brought together. Excited by their faith and the possibility of plunder, the *bedu* played a significant role in the development of the new religion: they formed the military promoters of Islam, both in the early days of Muhammad's own conquests and in later centuries.

Muhammad's gospel was received unsympathetically and he was forced to leave Mecca for Medina where his support grew. This allowed him to fight his way back into Mecca where the Muslims were then allowed to worship in the Ka'aba which has since become Islam's holiest shrine, a rectangular building which surrounds the Black Stone, a meteorite which had been an object of worship since pre-Islamic times. Having taken Mecca and Medina, the Muslims continued to conquer the peninsula by force of arms or in many cases by treaty and, by 632, when Muhammad died, Islam was the religion of most of the peninsula. After his death the process of conquest continued, under the leadership of a succession of Caliphs, but soon differences arose and the first major splits in Islam took place, which resulted in the major division of the Islamic community into Sunni (Orthodox) and Shi'a, which survives to the present.

By the Eighth Century, the centre of Islam had moved from the Hejaz to Damascus and Baghdad. Thereafter the bulk of the Arabian peninsula remained a backwater of little political interest to the various empires which fought for domination of the Middle East. Indeed, it served as a relatively safe refuge for a number of dissident movements (variously known as *Kharijites*) to evade the control of the central imperial authorities.

The Hejaz alone, because of its pilgrimage and trading interests, remained a centre of attention and benefited increasingly from the pilgrimage to the Holy Places of Islam. Pilgrims travelled to Mecca and Medina by sea and land, bringing with them goods and culture of many countries (within a century after its foundation, Islam had reached from China to Spain). Some of the pilgrims stayed and started trading with their former homes. Servicing and exploiting the pilgrims became a major economic activity and a reason for struggles to control the Hejaz.

Islam, unlike Christianity, has a highly developed legal and social code of behaviour. At the purely religious level, the devout Muslim is required to observe what are known as the five pillars of the faith which are: repetition of the testimony of faith, almsgiving, five daily prayers, the great fast (Ramadan), and the pilgrimage to Mecca and Medina. But equally important to the Muslim is the *shari'a* law which prescribes in great detail exactly how the Muslim is to conduct his life within the community, as Orthodox Islam is very much a religion of the community, not of the individual.

Ideally, the Islamic community (*Umma*) is a society of equals before God. Thus in its early forms, before the acquisition of empire, there was no attempt to form a hierarchy. Some forms of religious organisation have however developed over the centuries: they largely coincide with the principles of tribal structure in which Islam had its origins. In accordance with tribal structure, the community is headed by a leader chosen by consensus; this leader is supported by a council of elders. As the Islamic empire developed, political power became divorced from religious power; within the religious structure the original organisation of spiritual leader (*Imam* or *sheikh*) surrounded by a council of elders (*'Ulema*) has continued to survive and wield considerable influence on a local level in day to day affairs.

The *'Ulema* can best be described as the learned men who bear the function of interpreting and explaining the legal and religious requirements of Islam to their local community. In this function their role tends to be more juridical than theological, since orthodox Islam is largely determinist, ascribing all events and changes of fortune to the will of God.

One specialised role within religious organisation is that of the *qadi* who presides over the religious courts. Another important function of the religious community centred on the mosque is education, firstly through the *kuttab* where young children are taught to memorise the Koran, sometimes learning literacy in the process. In these *kuttab* the entire emphasis is placed on religious education to the exclusion of all else. A pupil who successfully memorises and recites the Koran, thus gaining the honorary title of sheikh, may pursue his religious education by becoming a disciple of a great *alim* and acquiring knowledge until he himself joins the *ulema*. Of course this process applies only to men as religious education was not considered important for women.

Life in the interior of Arabia in the early 18th. Century was much as it had been in previous centuries, with minute city states located around some oases, heavily fortified to afford protection to the local agricultural community in times of war, which were a regular feature of life. There was constant warfare between these towns as they attempted to extend the areas under their control, with the nomadic lineage groups of the dominant families in the towns carrying out the war effort. Since the balance of forces from one town to the next tended to be equal, these struggles were usually indecisive and continuous. The pattern of struggles between nomadic tribes was repeated in the struggles of the towns.

Wahhabism

The history of Islam is punctuated with a series of dissident movements which considered that the main orthodox trend deviated from the true faith. Many of them objected to Sunni Islam's tendency to accept and incorporate pre-Islamic rituals and called for a puritan return to the fundamentals, as they saw them, of the teachings of Muhammad.

In the small town of 'Uyaina, in the Najd, Muhammad Abdel Wahhab was born in 1703 in a family of *ulema*; from childhood he studied religion and by the age of twenty had made the pilgrimage to Mecca. He then continued his studies in Medina and travelled for a number of years visiting Basra, Baghdad, Isfahan and other places studying theology and becoming an *alim*. On his return to Najd about 1740 he started to preach the new faith.

Abdel Wahhab's doctrines were based on a return to the basic principles of Islam as preached by Muhammad, and excluded all the pre-Islamic rituals and beliefs which Islam had absorbed. Only the Koran, the Sunni law schools and the six books of tradition which were adopted during the first three centuries of Islam were acceptable as bases of the faith; all later interpretations were excluded. Absolute monotheism is one of the most fundamental principles of Wahhabism: all signs of polytheism on the part of Muslims are considered heretical, worse than ignorance and false religions, worse than unbelievers. Polytheism was to be recognised in the use of the name of prophets or saints in prayers, asking for intercession from anyone but Allah, in the building of mausoleums as religious centres and in the visiting of tombs, or the worship of idols. The Koran is the sole basis of knowledge and it is heretical to interpret it. The creed is based on an extremely strict moral code which involves obligatory public attendance to prayers, the payment of *zakat* (religious tax) on all profits, including those of trade, a total ban on smoking, and the counting of the names of Allah on the knuckles of the hand rather than on beads. Wahhabi mosques are built with the greatest simplicity, and include neither ornaments nor minaret.

When he started to preach his doctrine in his home town, Abdel Wahhab was confronted with hostility; he then moved to Dar'iya, a neighbouring town which was ruled by the Al Saud.* There he struck up an alliance with the ruler Muhammad Al Saud. This was to be a long lasting and successful alliance in which the Al Saud committed themselves to be the temporal defenders of the Wahhabi doctrine and to defend it by force of arms if necessary, which it was.

Armed with the strict ideology of Wahhabism and the military power of the Al Saud, the movement developed rapidly throughout Najd, where oasis after oasis was conquered and the local ruler impressed with the necessity for conversion. The alliance broke the political and military stalemate which had prevailed in the Najd, unifying it under a single rule and even expanding to both the Red Sea and the Gulf coasts. By 1806 most of the peninsula, including al Hasa, Bahrain, Kuwait, the Pirate coast and the Hejaz, had been conquered and more or less converted.

When the Al Saud ruled this vast area they divided it into 20 regions which were each headed by a governor chosen from among faithful Wahhabis. The governor collected the taxes which had been assessed by the ruler, and levied troops as needed for the war

---

* Al Saud is the generic name for the family of Saud.

effort. The governor was assisted by a *qadi* who was responsible for religious education and propaganda. In each conquered town a fort was built and garrisoned by trusted troops, and a simple Wahhabi mosque built where the new faith was preached. Major decisions were taken by the Al Saud ruler with the advice of his entourage — those leaders of conquered and allied tribes who paid him allegiance.

The occupation of the Hejaz and the deposition of the Grand Mufti of Mecca brought the Al Saud into confrontation with the Ottoman Empire. Although the Turks had been expelled from the Gulf coast in 1669, they retained an interest in the Hejaz as the centre of the world of Islam, and appointed the Grand Mufti of Mecca to supervise and control the activities of the Hashemite, who since the 4th. Century were the hereditary *sherifs* of Mecca, claiming direct descent from Muhammad. When the Al Saud went on to attack Syria and Mesopotamia, the Sultan in Constantinople asked Muhammad Ali, the ruler of Egypt, to reconquer the Hejaz and expel the Wahhabi fanatics on his behalf.

Muhammad Ali, while paying formal allegiance to the Sultans, was busy building his own empire. As it suited his plan for control of the trade routes and of the pilgrimage he launched an expedition against the Wahhabis in 1811. By 1814 the Egyptians were in control of the Hejaz and in 1818 finally defeated the Al Saud at their capital Dar'iya, which was totally destroyed and made uninhabitable for future generations by cutting down the palm trees. The ruler Abdullah al Saud was sent to Constantinople as a prisoner, where he was beheaded. This was considered by the Ottomans and the Egyptians to be the end of the Wahhabis who were seen as nothing more than another petty rebellion.

This was not the case: in the next twenty years' occupation the population, although relieved of religious oppression, resented the invading army which had been fed from the very limited local resources. When the Egyptians left, local rivalries between petty rulers developed throughout the region as they renewed their attempts to dominate the Najd. By 1843 Turki al Saud succeeded in reuniting the Najd under his rule. He was assasinated and succeeded by his son Faisal, whose expansion brought back the Egyptians, who re-asserted control in the late 1830s, but Faisal returned from his prison and continued the struggle while stories of his heroic escape spread among the population. Muhammad Ali's death in 1849 ended all direct Ottoman and Egyptian interference in the heart of the peninsula.

After Faisal's death in 1865, internal strife among the Al Saud meant that their control of the Najd diminished as they fought each other for the succession. A rival family, the Al Rashid from the Shammar tribe, whose capital was Ha'il, took advantage of these squabbles to win control of Najd and its capital Riyadh, and sent the remaining Al Saud into exile in Kuwait, where they lived under the protection of the ruler, Sheikh Mubarak. Thus by the turn of the century Najd was under the control of the Al Rashid, al Hasa and the Hejaz were under Ottoman control and in the surrounding states of the Gulf, British intervention was becoming more formalised and threatening as Kuwait, Bahrain, the Trucial states, Muscat and Oman, the South Arabian statelets and Aden Colony were under varying degrees of British control. Being interested in the trade routes to India and strategic positions, Britain had not tried to involve itself with the interior of the peninsula or its regime, and had only clashed with the Wahhabis when their advance threatened British interests in the early 19th. Century in Muscat.

The following chapter will discuss how this poverty-stricken and divided region, subject not only to the traditional regional and tribal differences and conflicts, but also the increasing pressures of imperialism and the manoeuvring for domination, became unified under one family able for the first time in the history of the peninsula to subordinate 'permanently' the challenge of other families and to develop the basis for a modern state.

CHAPTER II
THE FORMATION OF
SAUDI ARABIA

In the age of imperialism, the world was carved up into clear areas of colonialism or spheres of influence between the imperial powers. Borders were often for the first time clearly defined, owing more to the balance of power between the imperialist states than to the geography or populations of the regions concerned. The imperialist powers struggled to ensure control of the entire planet, either for directly economic or for political and strategic reasons.

The process of unification of Saudi Arabia took place therefore in conditions new to the peninsula: in the struggles between the tribes to dominate the region, alliances were a major factor, particularly those with the British or the Ottomans. In the first thirty years of the 20th. Century the area was finally united into a single political unit, culminating in the proclamation in 1934 of Abdel Aziz al Saud as King of Saudi Arabia, thus introducing the concept of monarchy into a formerly tribal structure. The present state of Saudi Arabia was formed out of the struggle between the tribes of the peninsula and as a response to the spread of British influence in the Middle East. While Britain had little interest in the desert, it found it imperative to determine borders with its protectorates in the Gulf, except through the deserts of Oman and Yemen, since these were at the time considered worthless. Throughout these years of imperialist turmoil, Abdel Aziz al Saud strengthened and increased his control over the Arabian peninsula.

The recovery of Najd and the first phase of expansion

The first step in the building of the Saudi Kingdom came in 1902 when Abdel Aziz, with the backing of of Sheikh Mubarak, the ruler of Kuwait, launched a surprise attack on Riyadh, the capital of Najd and recaptured it from the ibn Rashid. This victory provided a base from which to take the rest of Najd: the Saud family returned from exile, the town was fortified in anticipation of an assault from the Al Rashid (which did not immediately materialise) and Abdel Aziz went on to take control of the areas of Najd where the Al Rashid control was weakest — in the south. He obtained the allegiance of the Al Kharj, Aflaj and Wadi Dawasir areas, and conquered the Qahtan tribe.

In the following years a series of battles between the two forces took place, none of them decisive, but with Abdel Aziz winning the areas in the north of Najd. In 1906 his victory over the ibn Rashid in Qasim gave him control over the north and with it the whole of Najd. As a result of the negotiations arranged by Sheikh Murabak after the Qasim victory, the Turkish troops which had assisted ibn Rashid withdrew. The possibility of the Al Rashid regaining their position in the region was further diminished by the death of their leader Abdel Aziz ibn Rashid in the battle for Qasim. His death brought about a period of instability in the ruling group in Ha'il: between April 1906 and September 1908 three men ruled in succession, each coming to power by murdering his predecessor. Ousting the Al Rashid from Najd did not make Abdel Aziz's position secure: over the next few years his control over Qasim was challenged on a number of occasions both by local rulers and by whichever ibn Rashid happened to be in power. Uprisings took place throughout Najd, following the traditional tribal pattern. In 1908 a drought which was to last many years exacerbated Abdel Aziz's difficulties: inter-tribal raiding increased as the only means to supplement economic resources. Abdel Aziz's resources were depleted by drought, affecting the means to support his army and his ability to buy the loyalty of the tribal leaders. He could not easily contemplate expansion to solve the crisis, as Najd was surrounded by British or Turkish-dominated territory. To compound his problems, in 1910 a branch of the Al Saud rebelled against him: the rebellion failed but its leaders found refuge in Ottoman-protected Hejaz, at the invitation of Hussein ibn Ali of the Hashemite family, recently appointed Sharif of Mecca.

Abdel Aziz responded to the situation in two ways: first he

founded the political-religious-military organisation which became known as the Ikhwan (see p.21), thereby creating a reliable army whose prime allegiance was to himself. Secondly, he occupied al Hasa, the eastern region, which was then under Turkish control: in May 1913, after a surprise attack, the Turkish forces were escorted to the coast where they embarked. The annexation of al Hasa alleviated the economic problems of the Al Saud regime by giving it control of fishing and pearl diving resources on the Gulf Coast, and of the date producing oases of Hofuf and Qatif. Unlike Najd, al Hasa is populated mainly by Shi'a Muslims who resisted Wahhabism, and have to this present day remained culturally separate from the surrounding region. As Govenor of Al Hasa Abdel Aziz appointed a relative of his, ibn Jiluwi who ruled with a harshness which has remained notorious: the administration of the region has remained in the control of his descendants.

A strategic consequence of the conquest of al Hasa was that by expelling Turkish forces, Abdel Aziz came directly into conflict with the remnants of the Ottoman Empire. He therefore approached Britain, the Ottoman's competitor in the region, for protection. He also proposed to the Turks that he would recognise their suzerainty over his entire territory in return for recognition of his hereditary rule over Najd, and considerable autonomy. Neither the Turks nor the British responded with any rapidity as they were busily sharing out the area between themselves and had not decided where Saudi controlled Najd fitted into their plans. These inter-imperialist plans were, however, interrupted by the declaration of the First World War in which the two states were on opposite sides.

Arabia during the First World War

In the Hejaz the War seemed like a good opportunity for the Hashemites to further their interests. Although encouraged by the Turks to declare *jihad* (holy war) on the Western alliance, Hussein decided to see what, if anything, the British would offer him for his support, despite his formal relationship of subordination to the Ottomans. Hussein's ambition was to set up an independent Arab Kingdom, including vast areas of the Arab world, in which he would be King.

In November 1914 the British offered to guarantee his continuation in office and to support his right to help the Arab movement in return for support in the war. Seeing this as a starting point he continued negotiations to obtain a further commitment from Britain while he prepared his troops for what came to be known as the Arab Revolt.* In 1916 Hussein proclaimed himself King of the Arabs and claimed British support for this move although Britain only recognised him as King of the Hejaz.

In the Najd, the outbreak of hostilities completely transformed Abdel Aziz's position. Instead of seeking support, he suddenly found himself being asked for it. The Al Rashid, still in alliance with Turkey, attacked Abdel Aziz's forces in 1914. In early 1915 in his counter-attack, Abdel Aziz was accompanied for the first time by a British adviser, Captain W. Shakespear, who was killed in the incident. Later that year a treaty was signed at Qatif between Abdel Aziz and Percy Cox, Chief British Political Officer in Mesopotamia. By the terms of the treaty Britain recognised Abdel Aziz's independence as ruler of Najd, al Hasa, Qatif and Jubail and as absolute ruler of the tribes of the area; Britain guaranteed the integrity of this territory against foreign aggression. Abdel Aziz undertook not to maintain any relations with other foreign powers nor to alienate any part of his territory or give concessions without prior consent from Britain, nor to attack the British-protected principalities of the Gulf and to keep open the pilgrimage routes. In other words, the Saud were now bound to Britain by a treaty of protection similar to those of the other Gulf states.

The Treaty of Qatif remained extremely unclear about Najd's borders with the Hejaz: Abdel Aziz, aware of the close relations developing between Hussein of Mecca and the British, asked for clarification, which he did not get as Britain was at that time promising Hussein support for his ambitions. Instead Abdel Aziz was invited to Basra to receive decorations and during the ensuing negotiations, in late 1916, he was promised a monthly subsidy of £5,000, four machine guns and 3,000 rifles provided he kept a force of 4,000 men continually in action against ibn Rashid. This was partly meant to divert Abdel Aziz from concerning himself with the Hejaz, but it did not.

While the war was still going on, a clash took place in 1917 at the

---

* The Arab Revolt was an armed uprising against the Ottoman empire which took place in Hejaz and the Fertile Crescent during the First World War, under the leadership of the Hashemites and sponsored by Britain.

Khurma oasis between the Hashemite and the Wahhabi forces, with victory for the Wahhabis. This came as a surprise to the British, who had expected the Hashemites to win and had supported their claim. But they were now obliged to recognise, if only tacitly, Wahhabi control of the strategically located oasis.

Definition of the borders of Saudi Arabia

Although Abdel Aziz decided against the immediate invasion of the Hejaz, he felt threatened by the Hashemite manoeuvres to control Iraq, Syria and Transjordan, that is, all the states to the north of Najd. Success would give the Hashemites control of all the land connections of Najd with the rest of the world, leaving it totally dependent on sea communications. To prevent this encirclement, Abdel Aziz launched in 1918 an attack on the Al Rashid, then allied to the Hashemites, and in 1921 finally took their capital Ha'il. Abdel Aziz went on to extend his control northwards to Wadi Sirhan and Jawf, thus defining the northern borders of his domains to their present limits and establishing the domination of the Al Saud over the ibn Rashid.

Having defeated the ibn Rashid, Abdel Aziz turned once again to the Hejaz and in 1924 the Ikhwan attacked Ta'if and occupied it without difficulty. King Hussein was forced to abdicate and was succeeded by his son Ali who abandoned Mecca, giving Abdel Aziz the opportunity of making the pilgrimage there in 1924. By the end of 1925 Jeddah and Medina had also come under Al Saud control, and the rule of the Hashemites in Hejaz came to an abrupt end.

To establish his position in Islam and to show that he intended to rule Hejaz according to its laws, Abdel Aziz was proclaimed King of Hejaz in the Great Mosque of Mecca on 8 January 1926. This consecrated Wahhabi rule in the Hejaz whose population had already shown its hostility to the puritanism of the new rulers.

Between 1920 and 1933 the Al Saud struggled to obtain control of Asir from the Imam of Yemen. This was finally resolved in June 1934 when the Treaty of Ta'if was signed: this gave Saudi Arabia full control over Asir, including the town of Najran, and a war indemnity of £100,000 in gold. The border was defined by a border commission in 1936. Thereafter relations between the Imam of Yemen and Abdel Aziz improved and later became excellent.

Having settled his relationships with local rulers it remained necessary for Abdel Aziz to establish relations with Britain on a

THE FORMATION OF SAUDI ARABIA 19

new basis after his victory over the Hashemites. During the War and post-War period, British policy had been divided between support for the Al Saud in Najd and for Hussein in the Hejaz; this did not merely reflect different individual opinions, or the comparative abilities of the two leaders, but stemmed from the structure of British colonial administration. Abdel Aziz was dealt with in Najd by the officers based in Mesopotamia who were dependent on the India Office and reported to London via Bombay, while those who dealt with Hussein came from the Arab Bureau set up in Cairo under the control of the British Foreign Office. There was hostility not only between Abdel Aziz and Hussein, but also between their supporters in the British administration, H. St. John Philby and T.E. Lawrence respectively. Lawrence backed Hussein's demands for recognition of an independent Arab kingdom after the war, but failed to obtain assurances from the Foreign Office to this effect. Within the British administration in-fighting continued between the two factions after the war, but eventually the solution came from the situation on the ground, in the form of Abdel Aziz's victory over the Hashemites.

(By 1927) 'It was obvious enough to all concerned that the old 1915 Treaty of Qatif no longer represented the true relations between the two countries; and that a new treaty was now needed to provide British recognition of Ibn Sa'ud's complete sovereign independence, with all that that implies. He must be free to have relations with other Powers, and have the right to supply himself with arms and ammunition from any available source without restriction: while there could be no question of his recognising the old capitulatory regime, born under the Turks and continued with some modifications during the short Sharifian interregnum. But the British government clung tenaciously to its old right of manumitting runaway slaves who might take sanctuary in its Consulate...'[1]

It was from this standpoint that the Treaty of Jeddah was negotiated by Abdel Aziz and Sir Gilbert Clayton. Signed in May 1927, the treaty asserted Abdel Aziz's complete and absolute independence and declared void the Treaty of Qatif (1915); it provided for non-aggression and friendly relations. Abdel Aziz acknowledged the special British position in Bahrain and the Gulf Sheikhdoms and agreed not to attack them; he also agreed to

1. H. St. John Philby, *Sa'udi Arabia*, London 1955, p.302-3.

cooperate in the suppression of the slave trade. No British subsidy was stipulated in the Treaty which was valid for 7 years and was renewed in 1934.

The 1915 Treaty had marked formal dependence of the Al Saud on Britain. Unlike other Gulf states, however, Saudi Arabia succeeded in asserting its independence from Britain. The 1927 Treaty proved that Britain had not really dominated the Al Saud in the way it had been able to dominate the other states with which it had had a similar relationship. Abdel Aziz had successfully manouevred within the imperialist game and retained the independence of his newly formed country. There is no doubt that this would have been more difficult, if not impossible, had it then been known that it contained vast oil reserves.

But in 1926 one of Abdel Aziz's motives for conquering the Hejaz was his desperate need for the revenue afforded by the pilgrimage. The conquest did raise some problems at the religious level: the Wahhabis, and in particular the Ikhwan, were known throughout the Islamic world for their extreme puritanism and intolerance. Therefore, although there was dissatisfaction with Hashemite rule of the Hejaz by the mid-1920s, the fact that they were replaced by the Wahhabis in the Holy Places caused considerable concern among influential Muslims. The main Muslim community leaders tried to make life difficult for the new regime in a number of ways: the Iranians sent a commission of inquiry into the damage done to tombs in the sacred cities in the course of the Ikhwan onslaught. The Indians called for the formation of an international organisation representing all Muslim states to administer the holy cities. In response, Abdel Aziz announced that he would continue to rule Hejaz according to the laws of Islam and in the interests of the entire Islamic community, thus reasserting his political control. He also invited all Muslim leaders concerned with the future of the Holy Places to come to Mecca in the summer of 1926 when a conference would be held to discuss the organisation and administration of the *Hajj* (the pilgrimage), with the hope of increasing the number of pilgrims which had declined in recent years, causing heavy losses to the Treasury. Abdel Aziz was counting on the income from the *hajj* to sustain his kingdom.

The Islamic Congress was held in June 1926 in Mecca and was attended by over 70 major Islamic leaders. Abdel Aziz asserted his intention to retain control over the temporal administration of the

Hejaz and declared the Holy Places to be in trust for Islam as a whole. He proceeded by asking the congress's advice on how best to serve the religious needs of Islam. The congress adjourned after having taken certain decisions on pilgrim traffic and the improvement of sanitary conditions. It had given Abdel Aziz the recognition he sought from Islam for Wahhabi rule, and produced a friendly atmosphere despite another incident over the Egyptian escort of the *Mahmal*.* On this occasion the Ikhwan were roused by the misbehaviour of the Egyptian armed escort who were playing bugles and attacked them: 25 people were killed before Abdel Aziz intervened personally to prevent further bloodshed. As a result diplomatic relations between the Hejaz and Egypt were broken and only re-established in 1936, ten years later.

The Ikhwan Movement and its Significance

Accounts of the Ikhwan movement vary considerably both in substance and in detail, but there has been little discussion either of its origins or of what became of its members after the movement was crushed in 1930.

By 1911 Abdel Aziz had conquered most of Najd and was looking towards al Hasa and the Hejaz to solve his economic problems — the Najd being a poor area barely able to sustain its population at the best of times, let alone in the throes of the drought which began in 1908. In addition he realised that the 20th. Century was a period in which radically new forces were operating in the peninsula, as witnessed by British involvement in the Gulf and the pressures to define borders, a practice totally in contradiction with the traditions and needs of nomadic life.

The position of the ruler of these vast territories is not hard to recognise: Abdel Aziz needed to build up a solid and reliable military force which would be more stable than the temporary war levies which had previously been used and whose morale and reliability depended primarily on the amount of loot to be gained in battle. His expansionist plans demanded that an armed force be devised which combined the qualities of the town dwellers with the mobility of the *bedu*.

* The *Mahmal* is the litter on which the cover of the Ka'aba was ritually carried from Egypt every year at the time of the pilgrimage. It was traditionally followed by a large escort including musicians.

As John Habib explains:

'The villagers of Najd were the most loyal citizens and most reliable soldiers but they could not leave their fields and shops for extended military service in campaigns far from their homes; on the other hand, the bedouin in their primitive state were too opportunistic and fickle in their loyalties to provide the dependability which an Arabian leader required for distant long range conquests.'[2]

In addition, Abdel Aziz had not attempted to follow the recapture of Najd with any new forms of political organisation. There were still frequent uprisings and no reason to assume that they would cease, nor was it possible to conquer other areas with any expectation of long-term control. Like his ancestors before him, Abdel Aziz had operated according to the traditional forms of control in which the ruler of a defeated tribe paid tribute and *zakkat* to Abdel Aziz, made his men available for armed service and attended the *majlis* in Riyadh. It was easy for a leader to stop paying tribute and to disappear as soon as he had recovered enough strength to challenge the currently dominant ruler.

Finally the experience of the Wahhabi empire in the 19th Century had shown that the Wahhabi faith was not enough in itself to create the basis for a long-term political structure able to hold together a political unit larger than a tribe or a tribal section. Although the early Saudi conquests created religious unity throughout Najd and beyond, religion was not used to unite the area politically or administratively, nor was anything else. As a result Muhammad Ali's Egyptian army, operating according to modern military principles, defeated the Wahhabis by a combination of tactical skills and the manipulation of existing tribal divisions. The Al Saud conquerors had based their rule on the traditional tribal allegiance system and failed to create national political structures able to integrate the tribes into a single unit. This failure to adapt to the modern situation was one of the causes of the Wahhabi defeat and disintegration in the 19th. Century. The lesson was not lost on Abdel Aziz.

Both to pursue his expansionist aims and to ensure effective control over the conquered areas Abdel Aziz sought a way of

---

2. John S. Habib, *The Ikhwan movement of Najd: its rise, development and decline*, Ph.D Thesis, 1970, University of Michigan, USA. p.8.

unifying his forces in a structure which would cut across tribal allegiance and create a commitment to a common leader and objectives. The tool he created was the Ikhwan movement.

Wahhabism, the uniting flag of his ancestors, provided the basis on which to build. The faith was still alive in the minds of some: like that of early Islam it was geared to conversion and expansion. By cutting across tribal divisions it created a unifying factor, and it was a distinctive sect which could clearly identify those who subscribed to it from others. Wahhabism could also be expanded beyond religion to include political and social organisation.

To develop the Ikhwan movement, Abdel Aziz sent out into the areas under his control Wahhabi missionaries who concentrated their efforts on the *bedu*. He hoped to convert substantial numbers of nomads, settle them on agricultural land, and turn them into agricultural communities with a feeling for land ownership, whose primary allegiance would be to Allah and the Al Saud, rather than to their tribes of origin.

The first Ikhwan settlement or *hijra* (pl. *hujar*) was founded in 1912 by some Muti' tribesmen at Artawiyah, a well on the Kuwait-Qasim route. This settlement remained the most important as well as the most famous:

> 'Ibn Saud, who had started the process of regeneration among the tribes through his missionaries, placed all necessary facilities at their disposal: money, seed, and agricultural implements, religious teachers, and the wherewithal for building mosques, schools and dwellings: and last but not least, arms and ammunition for the defence of the faith, the basic article of which was the renunciation of all the heathen customs and practices of the old tribal code. The brotherhood of men who accepted the new order, regardless of their tribal affiliations and social status, canalised the warlike propensities of the Arabs in the service of God and his representative on earth. Inter-tribal raids, highway robbery, tobacco and all the other amenities of the old life became taboo; and all attention was concentrated in the colonies on preparation for the life hereafter.'[3]

Later another famous settlement, that at GhutGhut, organised by the 'Utayba leader ibn Bijad ibn Humayd, was founded. They were followed by many more. Supporters of the Saud family have

3. Philby, *op.cit.* p.261-2.

put the total number of *hujar* at over 200 but it is more likely to have been about 60. Their size at their peak period varied from a dozen members to over 10,000: the latter is an accepted figure for Artawiyah, but this was undoubtedly the most successful and largest *hijra*.

In the *hujar* nomads settled and adopted Ikhwan customs. Agriculture and trade were supposed to provide the economic base, subsidised by the regime with seeds, tools, money for the building of houses and mosques and the supply of ammunition and weapons: each *hijra* had its religious teacher to train the Ikhwan in true Islam, i.e. Wahhabism. All other Muslims were regarded as little better than unbelievers and some, particularly the Shi'a were more hated than Christians. Raiding, a basic *bedu* tradition as well as a necessity, was not forbidden but encouraged, though new rules applied: true Muslims and other Ikhwan were not to be raided, but unbelievers and bad Muslims were, for this was an execution of God's will:

> 'The geographical distribution of Ikhwan colonies, and the wide range of tribes which they represented, enabled the striking arm of Ibn Sa'ud to be flexed in such a way that no part of the peninsula would be more than a day's march from the wrath of his Ikhwan. The tribal distribution provided tribal, clan and family links to all the major tribes of Nadj. The synthesis of geographical and tribal considerations provided Ibn Sa'ud with a network of military type cantonments that at once served as an outpost of loyalty and a collection point of intelligence at the farthest points from Riyadh during peace; in war, they became centres of mobilisation and access to given targets. Ikhwan troops marching from the farthest corners of Najd, say from al Artawiyah, could find brothers in arms in the *hujar* of the Ikhwan in the Hijaz and Jawf, and in between these two points find *hujar* which would give them provisions, water, intelligence, and other essentials as they stopped there during the march. The *hujar* served as military bases, supply bases, and religious outposts, and since many of them were located close to the traditionally sedentary places — such as Ghut Ghut's location in relation to Mhazamiyyah — they acted as a disciplinary influence on these towns, keeping them well within the Wahhabi fold.'[4]

4. Habib, *op.cit.* p.112-3.

Though at first there was an attempt to develop the Ikhwan movement as non-tribal with settlements including people of different tribes, creating support for the regime without tribal commitment, this was unsuccessful and the *hujar* were all tribally based and their leaders traditional tribal leaders. The two main leaders were Faisal al Duwish of the Muti' who led the most important *hijra* at Artawiyah and Sultan ibn Bijad Humayd of the 'Utaybah who led the GhutGhut *hijra*.

The Ikhwan wore short robes and special headgarments, a white cordon instead of the usual black *aqal*; this made them immediately identifiable. They disapproved violently of any modern inventions and devices as well as of any contact with non-Wahhabis, let alone foreigners: the British officials Philby and Dickson relate how the Ikhwan clearly showed their distaste for them and other Europeans. They were heavily subsidised by the court. They were very fanatical and even in the early days of the movement, Abdel Aziz had to control this: their fanaticism was largely due to a recent conversion after generations of superficial religious involvement. They were very courageous, even reckless in battle: their determination to fight and win against all odds, and in the face of armoured cars or the British Royal Air Force, resulted in extremely high casualties. This did not deter them as they desired death as a means to get closer to Allah.

Their military tactics included night marches followed by short hit and run raids, during which all enemy men were killed. These tactics and their lack of fear inspired widespread dread well beyond the areas in which they operated: this alone caused the evacuation of Mecca at their approach after they had taken Ta'if.

These characteristics and the Ikhwan's unquestioned devotion to Abdel Aziz rapidly made them the most important movement in the creation of Saudi Arabia. Apart from being the main military force and therefore responsible for all the conquests from that of al Hasa in 1913 to that of the Hejaz in 1925, they were a unifying factor in their imposition of strict Wahhabism and its regulations wherever they went. This had positive and negative aspects. For example they tried forcibly and brutally to convert the inhabitants of al Hasa, who were traditionally Shi'a. For a long time the Shi'a were prevented from holding their religious meetings and even from smoking in their own homes; incidents in which Ikhwan beat them and even murdered them were reported. Abdel Aziz had to forbid forcible conversion as the situation became extremely tense in the area. On several occasions, severe punishments were needed to control the fanaticism of the Ikhwan.

There had been opposition to the Ikhwan movement from its inception, particularly from ibn Jiluwi, the governor of al Hasa, who expected it to bring trouble in his region, which it did. The Ikhwan also considered themselves to be superior to the ordinary population, even when these people were Wahhabis; they were particularly hostile to nomads.

A further source of conflict between the regime and the Ikhwan was found in the contradiction between Ikhwan belief in the most rigorous forms of puritanism and Abdel Aziz's compromises with the modern world. These took various forms: the introduction of modern communications, such as cars, radio, telephone and telegraph. There was also opposition to the treaties and agreements made with infidels defining borders, and as a consequence the restrictions put on *jihad* by the Ikhwan who wanted limitless expansion. Abdel Aziz forbade forcible conversion and went even further by honourably receiving Christians in his tent instead of murdering them. Once borders had been negotiated, Abdel Aziz was bound to forbid cross border raids: this limited the Ikhwan's income as they relied on loot from raids as their major source.

Trouble started between the Al Saud and the Ikhwan when the latter began to defy Abdel Aziz's authority: this happened quite early on over a number of internal issues, but it was only after border agreements had been signed that the Ikhwan's border raids caused Abdel Aziz international embarassment. During the conquest of the Hejaz, the Ikhwan felt snubbed by reproaches of brutality at Ta'if and the fact that they were not allowed to storm Mecca, Medina, or Jeddah. The incident over the *mahmal* (see p.21) further offended them as they considered it fully within their rights to prevent such an idolatrous activity being performed in the holiest place. Further offence was given by Abdel Aziz because he did not ban tobacco in the Hejaz.

In 1926 the Ikhwan met in Artawiyah and sent a number of demands to Abdel Aziz. As a result a conference was held in Riyadh where outstanding issues were discussed by the Ikhwan, the Ulema and other notables, and the Saud. Compromise was reached on some issues, for example, telephones were declared lawful but gramophones and cinema were not. The major confrontation was to come.

In the following year the situation worsened, partly as a result of clashes on the Iraq borders, in which the British RAF massacred the Ikhwan who had crossed the border on a punitive raid to demolish a border post which the British were building, thus breaking the terms of the border agreement reached with Abdel Aziz. These clashes and the positions taken by the respective parties further embittered the situation.

By 1929 the situation between Abdel Aziz and his supporters on the one hand and the Ikhwan and theirs on the other, had reached breaking point and a battle took place at Sibila in which the Ikhwan were defeated. In January 1930 a further battle marked the final defeat of the Ikhwan.

This defeat, which some observers consider was due to British cooperation with Abdel Aziz, effectively marked the end of the movement. Many *hujar* were destroyed, including that of GhutGhut, and others were gradually abandoned.

> 'Their defeat at the battle of Sbila rang the death knoll of the movement as a meaningful military power and as a vital force guiding the development of the new state. The Ikhwan rebellion was the final effort of fanatical bedouins led by traditional tribal chiefs to continue their influence and control over the new, rising nation... The rank and file remained on their *hujar* collecting their subsidies but their direct influence on the affairs of state had paled. They were no longer the dominant military power in the country. Yet their ascetic existence, their extreme attention to religious detail and their high-spirited nationalism continued to influence the character of the nation which they helped to create and bolster.'[5]

An example of how the Al Saud regime used what remained of the Ikhwan movement can be seen in a revolt which took place in 1932: there for the first time they were used to preserve and defend recognised Saudi territory, rather than to conquer new ones, and they were led by Saudi leaders rather than their own traditional tribal chiefs.

A number of questions concerning the Ikhwan movement remain unanswered. First it is unclear to what extent the movement as such

---

5. Habib, *op.cit.* pp. 9-10.

developed beyond its tribal origins: for example its main leaders like al Duwish remained tribal leaders throughout; further he not only rebelled as a member of the Ikhwan in the late twenties, but was already in rebellion against the Al Saud in 1910, before the movement was created, and clearly remained loyal only while it suited his purpose. The movement certainly failed to overcome tribal divisions; the reason for this is to be found in its origins. The second problem relates to the actual number of *hujar* which existed: it is clear that settlements like GhutGhut and Artawiyah existed (the first was destroyed after the rebellion and the second still exists, as it was visited by Habib). How many others there were, and how many Ikhwan there were is far less clear: estimates vary between 60 and 200 settlements each including anything from 10 families to 10,000 inhabitants.

Third, there is no account of what happened to the Ikhwan after the end of the rebellion: did they return to nomadism? or did they remain agriculturalists? It seems likely that some of them joined what became the National Guard and that others, whose religious devotion was greater than their military interest, became members of Committees to Promote Good and Suppress Evil, later renamed Public Morality Committees, and otherwise known as the Religious Police (they make sure that Wahhabism is properly respected, prayer times followed, etc...). Although Habib does not give details of how he came to that conclusion, he claims they are now loyal members of government service:

> 'Once they ceased to be a military-political threat to the regime, they gradually worked their way into the confidence of the government. In addition to bearing arms, and forming the backbone of a home militia (the irregular units of the White Army), they and their children were given positions of trust and responsibility in the ever expanding government, since they were regarded as the most stalwart citizens, firmly committed to Islam as an ideology and to the patriarchal-king system of government as the appropriate form of rule. In return for service and allegiance to the King, they demanded, and received, unhampered access to him and his court. And this tradition persists to the present day.'[6]

Saudi Arabia in 1934

By 1932, Abdel Aziz felt secure enough in his control of the

6. Habib, *op.cit.* p.300.

Kingdom of Hejaz, Najd and its Dependencies to rename it the Kingdom of Saudi Arabia and have himself proclaimed King.

By 1934, having defined the borders of his Kingdom, Abdel Aziz was faced with the problem of unifying, or rather creating, its administration. Having occupied the Hejaz where a number of foreign legations were settled, the regime had to establish some identifiable institutions. In 1927 the USSR, the UK, France and the Netherlands had recognised the Wahhabi regime in the Hejaz. In 1925 the Directorate of Foreign Affairs had been set up and in 1930 this became the Ministry of Foreign Affairs under Prince Faisal. To this day the Ministry of Foreign Affairs is in Jeddah. The Ministry of Finance was created in 1919 under Abdullah Sulaiman al Hamdan who ruled it like a private fund, spending freely without referring back to the King, let alone anyone else.

Other decisions were taken by the King on consultation with his advisors who were mainly from other Arab countries, they included Yusuf Yasin, a Syrian and his political secretary, and Hafez Wahba, an Egyptian who was responsible for education and in this capacity opened a few boy's schools taught by Egyptians and following Egyptian curricula.

The process of forming a nation was not dealt with. From the beginning Abdel Aziz failed to enforce Wahhabi rules in the Hejaz, while they were being enforced elsewhere: for example tobacco was permitted there as the King was made aware of the revenue he could draw from it. Provinces were ruled separately and according to traditional principles of allegiance, rather than on the basis of a central administration. Local governors were usually members of the Saud family or of related or allied families such as the ibn Jiluwi in al Hasa, or the Sudairi in Asir. Tribal chiefs continued to be paid subsidies, which were called compensation for the ban on inter-tribal raiding. Within local government structures, tribal hierarchies were adhered to, and no attempt at modernisation was made.

There was no physical infrastructure to speak of: the Damascus-Medina railway had been put out of action in 1916, and has not been restored, there were no paved roads, the royal cars were driven in the open desert, ensuring their rapid disintegration. Apart from the radio network to allow members of the royal family to communicate between Jeddah, Riyadh and Ta'if, there was only a weekly paper, which became the state gazette later on, but began publication in Mecca in 1924 under the title *Umm al Qura*. Industry was non-existent, there were only a few traditional handicrafts; pearl diving had suffered a mortal blow with the development of Japanese cultured pearls. In Jeddah there were a few foreign banks and trading houses.

Saudi Arabia's main problem was economic. Shortage of resources had plagued Abdel Aziz from the earliest days of his political career. State income was meagre: it had risen from about £50,000 a year in the first decade of the century to twice that amount in the second, after which a British subsidy of £5,000 a month (from 1916-1924) must be included. By the thirties, the conquest of the Hejaz and its pilgrimage and trading resources brought the annual income up to £5 million. The number of pilgrims rose to 130,000 in 1927, the highest number recorded, but soon after that fell as a result of the world slump, and by 1931 only 40,000 Muslims made the pilgrimage. This fall resulted in a heavy loss of revenue, only partly made up by foreign aid from the USSR.

Foreign aid was not enough to solve the problem. The total absence of budgeting combined with careless spending meant that the economic crisis could only get worse. Government employees went unpaid for months (the drivers of the royal fleet of cars went on strike in 1931 as a result of non-payment, in the first ever strike in the country: they were all beaten up and sent back to their countries of origin). Suppliers also remained unpaid, whether for essential or luxury goods.

The crisis called for consistent efforts to create new sources of revenue. The traditional political economy gave no opportunity for the development of economic resources by the local people, as there was no obvious base for the development of a productive sector able to create a surplus. The finances needed by the regime had in the past come from abroad, like the British subsidy, and the King's reaction to crisis was to try and find other ways of obtaining income from abroad. As part of this process, he granted an oil concession in 1923 to Holmes, a prospector from New Zealand, who was to pay £2,000 a year, but this soon lapsed. A gold mining concession was signed in 1934 for a mine which was worked until 1954, when it was deemed unprofitable.

Negotiations took place between 1930 and 1933 for an oil exploration concession. In 1933 the concession was granted to an American company at the cost of £50,000 gold down payment, a sum which helped solve the regime's immediate problem. It was in this state of limbo that Saudi Arabia staggered through the 1930s unaware that the £50,000 gold which seemed like a bonanza in 1933, was merely the first trickle of a deluge which would emerge with the discovery and production of oil in the following decades.

CHAPTER III
SAUDI ARABIAN OIL:
FROM DISCOVERY
TO INTERNATIONAL POLITICS

By the early 1930's oil had become a major source of world energy: ships were oil-powered, the car industry was growing fast, and prospects for expansion seemed limitless. The competition war between the oil companies had been settled by the Achnacarry price fixing agreement of 1928, which ensured maximum profits for the big seven, five of which were US owned (Standard Oil of New Jersey, Gulf, Texaco, Socony Mobil, and Standard Oil of California), one British (Anglo-Persian, later British Petroleum) and the seventh half British half Dutch (Royal Dutch Shell). The race for new oil concessions was a basic activity for the major oil companies. In the Middle East the Red Line Agreement of 1927 had defined those areas in which the various companies could obtain concessions. This Agreement included the Arabian peninsula but excluded Kuwait as a preserve of the Iraq Petroleum Company (IPC). Because of this agreement, when Standard Oil of California (SOCAL) obtained a concession in Bahrain, it had to masquerade as a British company by registering in Canada as the Bahrain Petroleum Company (BAPCO).

A similar manoeuvre was not necessary in Saudi Arabia where Abdel Aziz's military victories had forced Britain in 1927 to relinquish its 1915 protectorate treaty and to recognise the full independence of the Al Saud regime. Moreover, in 1923 the British government had agreed to a petroleum concession being given to a non-Briton, Holmes, who although connected with a British company, had already offered and sold his Bahrain interests to the Americans. His concession in Arabia lapsed four years later. Therefore when SOCAL sought a concession, there was no reference to the Red Line Agreement even though there was direct competition between the British controlled IPC and the American SOCAL, which was not a party to the Agreement.

SOCAL's first approach to the King in 1930 was rejected. However, after its success in Bahrain, the company tried again and negotiations took place in Jeddah in 1933, where the representatives of competing companies assembled. Although Holmes had reappeared on the scene, he was not a serious competitor, but IPC, represented by S.H. Longrigg, was there, taking a traditionally paternalistic colonial attitude and acting as if Saudi Arabia, being within the British sphere of influence, would not dare reject a British offer. The company did not take into account the King's independence of mind, nor the threat to his rule posed by his perennial financial difficulties. British interest in the concession was limited to preventing other companies exploiting the oil, which would increase production and lower world prices. They were not themselves interested in exploiting the concession at that time. IPC was also unaware, at the time, of the serious long-term threat which US control of Saudi oil resources might pose for Britain's position in the world economy.

When asked by the Saudi government for £100,000 gold, the IPC offered £10,000. SOCAL offered Abdel Aziz £50,000 gold, got the concession, and has been laughing all the way to the bank ever since. As Tanzer put it:

"For this amount Standard of California got a sixty-year concession for all of Saudi Arabia's oil — a far better bargain than even the fabled deal giving the American Indians $24 for Manhattan island. By 1936 Saudi Arabia looked so promising that Standard of California, which had no outlets for vast quantities of crude oil, sold a halfshare of its Bahrein and Saudi properties to Texaco for $3 million down and $18 million out of future earnings — another great bargain, this time for Texaco. The partnership between the two companies is known as Caltex."[1]

1. M. Tanzer, *The Energy Crisis* Monthly Review Press 1974 p.45.

The concession was extremely favourable to the company. In an area of 350,000 square miles in eastern Arabia, including coastal waters,

> "the company was given the exclusive right to explore, prospect, drill for, extract, manufacture, transport and export all oil produced... The company agreed to build a refinery; supply the government with 200,000 gallons of gasoline and 100,000 gallons of kerosene yearly; and advance loans deductible from future royalites, which were fixed at 4 gold shillings per ton of crude oil."[2]

West of this exclusive area was also a preferential area in which the company merely had to equal the offer of any competing bidder to obtain the concession: in 1939 a supplemental agreement was signed, enlarging the concession area by a further 100,000 square miles including a halfshare of the Saudi-Kuwait neutral zones.

The absurdity of these terms must be seen in the context of the period: oil concessions were totally in the hands of British and US companies who were expanding into Middle Eastern oil more as a strategic move than out of immediate need, as at that time most of the American oil companies' profits were made in the US. In Saudi Arabia, competition between British and US companies was possible because the American companies involved had not signed the Red Line Agreement. The British interest was limited to a concern that potential Saudi oil should not compete with their own Iraqi and Iranian sources. From the Saudi point of view, the concession did not appear as outrageously exploitative as it in fact was. It must be remembered that the country was so poor that the sums negotiated represented enormous wealth and that the ruling family had no conception of the profits that might be made from oil. The regime's attitude was dictated simply by its immediate financial crisis and, in addition, it was in a position of weakness in a world dominated by the imperialist powers.

The first years of oil exploration are remembered by Aramco employees as pioneering times comparable to the gold rush. Exploration was begun in the autumn of 1933 bringing to al Hasa large quantities of modern machinery and a number of Americans. They were compelled to wear traditional Arab clothing and were accompanied in their travels by an escort of Saudi armed soldiers to

---

2. N.Walpole and others, *Area Handbook for Saudia Arabia* American University, Washington, D.C. 1971, pp.244-5.

protect them from the hostility of the inhabitants of the eastern region. One of their first moves was to establish a relationship with one of the more prominent merchant families in the area, the Al Gosaibi, who acted as agent for their local needs. It is interesting to note that involvement of the local community in oil exploration was minimal and all the company's needs, ranging from housing materials to frozen chickens were from the first imported from the USA. The local fishing villages, which have now become the modern and westernised towns in the country, were at first hardly affected by oil exploration.

Oil was discovered in commercial quantities in 1938 in Dhahran and production began on a small scale the same year after a small storage and shipping terminal had been built in al-Khobar, making it possible to ship the oil to the refinery in Bahrain. On 1 May 1939, the first tanker was loaded to export oil, an occasion celebrated by a visit by the King with a modest retinue of 2,000 people. Ras Tanura was selected as a main terminal site and, despite the outbreak of war, work continued on this project which included a T-shaped pier with tanker berths, a submarine pipeline to the Bapco refinery in Bahrain, a refinery for 50,000 barrels a day, with storage tanks and loading lines. Most of these were built by 1945.

USA and UK struggle over oil resources during World War Two.
Despite the discovery of new oil fields, the war did slow down exploration: supplies were difficult to obtain because of shipping restrictions, and the number of American personnel was reduced dramatically after insignificant Italian bombing of Dhahran in October 1940. Caltex reduced its operations to a minimum level and, although output increased during the war, it was only with the ending of hostilities that a rapid and sustained increase in oil production got under way.

By 1940 it was clear that Saudi Arabia held vast oil reserves. It was this, rather than the regime's somewhat symbolic support of the allies, which had a considerable effect on Saudi Arabia as on the rest of the Middle East. The imperial conflict between the USA and Britain for post-war control of oil was fought out during the war, with the US emerging as the clear victor.

As in the First World War, Saudi Arabia was in the midst of economic crisis: the income from the pilgrimage was reduced almost to nil, and once again there was a serious drought which affected agriculture. The regime needed $10 million a year to keep up its extravagant standards and to pay essential subsidies to tribal leaders. In 1941, the King asked Caltex for a $6 million advance on royalties. The company was very sensitive to the threat posed to the regime by the economic crisis and fully supported the claim. It had no wish, however, to pay out of its own pocket, and attempted to persuade the US government to take on the commitment as part of its lend-lease programme. This necessitated that Saudi Arabia be declared a country whose defence was vital to the USA: at that time efforts to obtain this declaration failed, but the administration, aware of the importance of controlling Saudi oil, persuaded Britain to pay the sum out of its $425 million lend-lease loan, thus giving Britain the opportunity of appearing to be helping Saudi Arabia in its hour of need. At that time, the US government took so little interest in developments in Saudi Arabia that it did not even bother to maintain a permanent mission in Jeddah, leaving matters either in the hands of the Caltex representative or in those of the US Government's Cairo representative.

As the inter-imperialist struggle over oil resources intensified, it became very inconvenient for the US to channel its aid to Saudi Arabia through Britain, and in 1943 steps were taken to remedy the situation. Discussions between the oil companies and the State Department brought about the desired declaration from President Roosevelt in February 1943:

"in order to enable you to arrange lend-lease aid to the government of Saudi Arabia, I hereby find that the defense of Saudi Arabia is vital to the defense of the United States"[3]

By this time the British were escalating their involvement in Saudi Arabia by sending a military mission accompanied by geologists, and were trying to open a bank to issue paper money. This caused alarm in the USA and the Petroleum Administrator for War, one Harold L. Ickes, urged his government to buy up Caltex's Arabian operations. This attempt at nationalisation brought about the expected outcry from the oil companies, those dedicated supporters of free enterprise, and the proposal was reduced from 100 per cent ownership to 50 per cent to 33 per cent and finally came to nothing when it was defeated in Congress.

Abdel Aziz exploited the conflict between Britain and the USA to obtain maximum benefit and by 1946 had obtained aid totalling

---

3. *Foreign Relations of the United States*, 1944, V, p.734-6.

$17.5 million. It was also after 1942 that diplomatic relations between the USA and Saudi Arabia increased. A US legation was opened in Jeddah in 1942 and a US representative sent to keep a watchful eye on British activities in the country. Negotiations began for the construction of the US air base in Dhahran which was to be a link between Cairo and Karachi. It was also meant to ensure that the US "particularly [their] navy would have access to some of King Ibn Saud's oil".[4] Building was completed in 1946 giving the US their air base closest to Soviet industrial areas just at the beginning of the cold war.

As the war drew to a close, US supremacy over Britain in Saudi Arabia became clear. The struggle between the two had forced the US government to take some interest in internal developments in Saudi Arabia and to make some commitment to assisting the Saudi regime. This was meant both to protect Aramco (Arabian American Oil Company, the name adopted by Caltex in 1944) and to ensure the survival of a pro-western conservative regime in a world where opposition to imperialism was gathering momentum. Cooperation between the Saudi and the US government was to increase and reach new heights in the coming years.

The post-war period: Saudi Arabia gets a slightly better deal
The postwar period involved rapid expansion for Aramco: more oil fields were discovered, production facilities and the ports were expanded, and the Aramco city developed apace. Oil production rose from 20,000 barrels per day (b/d) in 1943 to over 500,000 b/d in early 1949.

As a result of its new supremacy over oil markets, the US now hoped to sell to the largest market, Europe, as at that time the US was self-sufficient in oil. To make this easier, in 1950 Aramco built a pipeline from its producing region to the Mediterranean to Sidon, which reached a capacity of 480,000 b/d and became known as Tapline. The cost of $230 million, was recovered within three years. To raise the capital for the operation, Aramco had to include new partners, changing its shareholding to the following: Socal 30 per cent, Texaco 30 per cent, Standard (NJ) 30 per cent, and Mobil 10 per cent.

With the vast expansion in Aramco production and profits, the income of the Saudi government also grew. But the royalty was

---

4. Quoted in G. Kolko *The Politics of War*, New York 1968, p.306.

38 A HOUSE BUILT ON SAND

still only 21 cents per barrel, while the company's profit was $1.10. Despite its increasing income, the Saudi government remained in permanent financial difficulties, with uncontrolled spending and no budget. This situation affected the regime's relations with Aramco, since it sought a solution by demanding a bigger share of the company's vast profits. This was solved by the 1950 so-called 50-50 profit sharing agreement, which involved an increase of revenue for Saudi Arabia, but no loss for Aramco. This miracle was achieved by calling the increased royalty an 'income tax' which, under the US tax code, could be deducted from tax payments at home:

> "Thus Aramco payments to the Saudi throne jumped from $39.2 million in 1949 to $111.7 million in 1950. Aramco taxes to the US Treasury were reduced by a similar amount to nearly zero, where they have remained ever since."[5]

US government support for such a deal shows the extent to which it was willing to support the oil companies. In 1974 George McGhee, then Assistant Secretary of State responsible for Middle Eastern Affairs, explained:

> "The ownership of this oil concession was a valuable asset for our country. Here was the most prospective oil area in the world. Every expert who had ever looked at it had said that this was the 'Jackpot' of world oil. To have American companies owning the concession there was a great advantage for our country...
>
> "We felt it exceedingly important from the standpoint of the stability of the regimes in the area and the security of the Middle East as a whole and the continued ownership of our oil concessions there and the ability to exploit them, that the government of Saudi Arabia receive an increased oil income."[6]

He did not bother to explain why the US Treasury took a decision whose

> "impact on the national treasury was direct and dramatic... the effect of the decision was to transfer $50 million out of the U.S. Treasury and into the Arabian treasury."[7]

5. Joe Stork, *Middle East Oil and the Energy Crisis*, Monthly Review Press, 1975, p.47.

6. *ibid.*, pp.48-9.

7. *ibid.*, p.48.

But he did make clear that the long-term objective was to assist US oil companies to control oil supplies to Europe and Japan, the US's competitors in manufacturing.

Further discussions took place throughout the Fifties aimed at providing the Saudi government with a fairer share of the profits, as the company was manipulating its price structure to ensure that the bulk of profits remained in its coffers. The company did this by basing its payments to the government on its realised prices, i.e. the prices at which it sold the oil to its own constituent companies, rather than on the published prices which were higher. In this way the company increased its already massive profits, which were carefully concealed. The constitutent companies in their turn traded the oil at large profits and without levy.

In 1953 King Abdel Aziz was succeeded by his son Saud. Shortly after his accession, King Saud tried to establish his regime's independence from Aramco, by creating the *Saudi Arabian Tankers Company,* a joint project with the Greek shipping magnate Onassis. Announced in February 1954, the plan involved the establishment of a fleet of tankers which would have the monopoly on transport of all Saudi oil not currently shipped by Aramco's own tankers; in the long run, Aramco would not be allowed to replace its old tankers, thus giving the Saudi company complete control of shipping its oil. Predictably, the announcement caused a storm of protest, not only in the US but also in Western Europe where shipping companies, who benefitted from the transportation of Saudi oil were based. The struggle over the issue continued in various national and international courts until 1958, when Aramco won its case at the International Arbitration Tribunal in Geneva, and the Saudis were compelled to abandon the project of an independent oil-related operation. Aramco had shown its muscle and maintained its control.

In another attempt to find better terms, the Saudi government negotiated concessions on territories which were either outside the Aramco concession or had been relinquished. In 1949 a 60 year concession was granted to Getty Oil which offered better royalties and a share of profits, as well as government participation in any future operating company. In 1975 the Japanese owned Arabian Oil Company was granted a concession which marked a significant improvement in favour of the state: the Saudi government was to receive 56 per cent of net income on all activities of an integrated company (i.e. a company including all stages of oil production, distribution, and marketing). One third of the seats on the board of governors of the company went to Saudi Arabians. Both these companies discovered oil and are now producing.

The birth of OPEC

By the late 1950s discontent among oil producing states rose dramatically as the oil companies unilaterally reduced prices, under the excuse of low demand for oil on the world market. The producers tried to get together to fight back. This took place at a time when there was a struggle for power in Saudi Arabia between King Saud who had brought his country to total bankruptcy thanks to his extravangance and lack of budgeting, and his brother prince Faisal who had a greater understanding of the need for a higher level of financial administration. The first US-trained Saudi Arabian oil technocrats were just beginning to start work for the royal house: most of them had been employees of Aramco's and were sent for training in the US at the insistence of the Saudi Arabian regime. Although in 1953 there were only 11 Saudi Arabians among the senior staff, the ratio rose later in the decade, and a number of men who had been trained in the USA returned to work in the state apparatus. One of them was Abdullah Tariqi who, at the age of 29, became the First Director General of Petroleum and Mineral Affairs at the Ministry of Finance and National Economy in 1954.

As the person in charge of government-Aramco relations, Tariqi's basic objective was to improve the Saudi negotiating position and, in the long run, to reduce and eliminate dependence of the Arabs on the oil company. He published a weekly newsletter to inform Arabs of oil matters as the companies' control of the industry had made sure of considerable public mystification on the problems involved. Tariqi negotiated the improved oil concessions with smaller companies, such as the Arabian Oil Company, but his main task was to concentrate on the relationship with Aramco. Here his aims were the following: increased Saudi Arabian participation at a higher level, encouraging the company to train more Saudi Arabians for high positions; pressure to turn Aramco into an integrated company and not merely the extraction vehicle of its constituents, thus giving Saudis the opportunity to gain experience at all levels of the oil industry from exploration to marketing. In the long run he sought to give Saudis control of all oil-related facilities; he also deplored the waste of resources as Aramco was allowed to flare gas without restriction.

From 1957 onwards, Tariqi tried to formulate a joint Arab policy on oil through the Arab League's economic council. His objective was to set up Arab-owned facilities such as a pipe line, a tanker fleet, a refinery and distribution network. This led to the First Arab Oil Congress in 1959 which was attended by Arab oil producers as well as observers from other oil producing states such as Iran and Venezuela. Debate led to the creation of the Organisation of Petroleum Exporting Countries (OPEC) the following year and marked the first public criticisms of the oil companies' secrecy concerning their operations and profits. A former employee of Aramco who worked as legal advisor to Saudi Arabia on oil matters argued that sovereign states had the right to alter contracts: "Any concession between a government and a company is not worth a damn if it does not please the people." The companies wanted to put their concessions "on a pedestal and wrap them up in their own flags to keep them away from political and economic developments in countries where they hold concessions. The purpose for which governments exist — service of their peoples — required that on proper occasions those governments be released from, or able to overrule their contracts and obligations."[8] Further attacks on the operations of the oil companies came a year later in the Second Arab Oil Congress, when the companies were directly accused of concealing their profits and of treating the producing states 'like children'.

The creation of OPEC in 1960 with the initial membership of Saudi Arabia, Iran, Iraq, Kuwait and Venezuela, was greeted with hostility by the developed capitalist world and the oil companies. They were horrified by the creation of a united front of the producing states whose overall aim was to defend their economic interests from the power of the oil companies, and whose immediate policy was to prevent any further reduction in oil prices, therefore in income, and to restore and increase prices to their former levels or higher. The companies refused to negotiate with OPEC and pretended that the organisation did not exist: they were able to persist in this arrogant attitude for a number of years. OPEC was in a weak position and, during its first years of

---

8. Harvey O'Connor, *World Crisis in Oil*, Elek Books, 1962 p.340.

operation, simply tried to persuade the companies that its position was correct, which it naturally failed to do. Producing states' oil revenue fell from 81.1 cents per barrel in 1955 to only 75.8 cents per barrel in 1962, when OPEC's activities succeeded in halting the fall: although posted prices continued to fall, OPEC, through higher taxes and royalties ensured that the producing states' income did not decline further, and by 1966 it had returned to the 1955 level.

In 1960 Tariqi became the first Saudi Arabian Oil Minister, but was ousted in 1962 when Prince Faisal finally gained control of politics in the country, and was replaced by Sheikh Ahmad Zaki Yamani, another US-trained technocrat. The change-over was attributed to Tariqi's increasing radicalisation and his Arabian nationalist approach to oil. His policies were in direct contradiction with those of the oil companies, and conflict between them could only have increased. Prince Faisal had no desire to start on a collision course with the USA as his policies called for cooperation with the States. Tariqi left Saudi Arabia and became an oil consultant; it is worth noting that he was not the extremist that he has been described as, and did not at that time call for outright nationalisation of the company, which he considered an unrealistic move.

Under the new Oil Minister, Saudi Arabia continued to try and sell concessions which would be more favourable to itself: in 1965 the French owned Société Auxiliaire de la Régie Autonome des Pétroles (AUXIRAP) obtained a concession which reflected the new terms OPEC countries were prepared to give: the government would receive higher royalties from oil, higher surface rentals throughout the period of agreement, and equity participation for the Saudi National Oil Company (Petromin) through an integrated company, should oil be exploited commercially. This concession has not yet produced any commercial quantities of oil.

OPEC gets results

In the late 1960s, the situation which had previously favoured the companies and the consumer countries suddenly changed as a result of an increase in world demand beyond the immediate distribution facilities (tankers and pipelines) and the inadequate refinery capacity. This came at the end of a decade when oil had been so cheap and plentiful, as to undermine the production, sale, and use of other sources of energy to the point where western industry was technically unable to cope with the temporary shortage. The full benefit of this shift in the balance of power was felt in the 1970s when the producing states were at last able to try and get an economic price for their oil. This was achieved through OPEC which, although it was by no means a unified organisation, struggled for a significant improvement in oil revenues for the producing states.

Although all producing states wanted increased income, the lead in the struggle was taken by the new revolutionary government in Libya which came to power in 1969. By concentrating its attacks on the smaller independent oil companies who had few, if any, other sources of oil, and by taking a slow and gradual approach, the Libyan government successfully raised the price of their oil by a substantial amount. This lead was followed by the OPEC Teheran agreement of 1971. It was not seriously contested by the companies who by then had to accept that the producer states had some rights and that the balance of power was rapidly changing. The Teheran agreement and subsequent minor adjustments reflected the changed situation:

> "From a market low of perhaps $1.25 per barrel in 1969 — of which about 10 cents was cost, 95 cents government taxes, and 20 cents company profits — by the middle of 1973 the market price had risen to about $2.50 per barrel, with $1.50 for the government and 80 cents for the companies. Both parties had gained significantly here, but while the government's per barrel revenues had increased by about three fifths, the companies' had quadrupled."[9]

Further improvements had to wait till the October War in 1973 which indirectly made the states independent of the oil companies in oil price decisions. In October 73, for the first time, OPEC unilaterally increased the price from $3.01 per barred to $5.12 for Arabian Light Crude, giving an increase in government revenue from $1.77 to $3.05. This increase was followed in December 73 by a further rise as a result of a number of oil auctions organised by the producing states, bringing the posted price to $11.65, giving the government $7 per barrel. The price of oil in early 1974 was 470 per cent above that a year earlier. In the case of Saudi Arabia, government revenue increased from about $1.2 billion in 1970 to over $26 billion in 1974.

9. M. Tanzer, *op. cit.* p.125.

Throughout this period of struggle, Saudi Arabian oil production increased. Aramco expanded its exploration, production and facilities although it was supposed to be passing under total Saudi control (see below p.46).

Crude oil production of Saudi Arabia (million barrels)

Year	Aramco	Total
1938	0.5	0.5
1940	5.1	5.1
1945	21.3	21.3
1950	199.5	199.5
1955	352.2	356.6
1960	456.5	481.4
1965	739.1	804.8
1970	1,295.3	1,386.3
1971	1,641.6	1,740.6
1972	2,098.4	2,201.9
1973	2,677.1	2,772.6
1974	2,996.5	3,095.1
1976	3,053.8	3,140.1

Source: Knauerhause p. 194, MEES 21.2.77

Daily production rose from 1.2 million barrels a day in 1960 to 8.2 in 1974 and 8.3 in 1976. Payments to the Saudi government also increased, though they cannot be compared with the profits of the oil companies, figures of which are obscured through their accountancy practice. While profits have risen and increasing quantities of oil have been extracted to little benefit for the producing country, it seems that published Saudi Arabian oil reserves are still rising.

Saudi Arabia's oil policies

Given that Saudi Arabia is the first exporter of oil in the world, it is important to discuss its role in the oil market both in terms of world oil politics where it plays a major role, and in terms of the consequences of these policies for the future of Saudi Arabia itself.

SAUDI ARABIAN GOVERNMENT OIL REVENUES bUS $ m.)

Year	Aramco	Total
1940	2.5	2.5
1950	56.7	56.7
1960	312.8	333.7
1965	618.7	664.1
1970	1,148.4	1,214.0
1971	1,806.4	1,884.9
1972	2,643.2	2,744.6
1973	4,195.0	4,340.0
1974	22,375.0	22,573.5
1975	24,838.6	25,676.2
1976		26,923.5[1]
1977		40,000*

* Estimate [1] MEES 4.4.77
Source: SAMA Statistical Summary 1977

SAUDI ARABIAN OIL RESERVES (billion US barrels)

Year	Year Aramco	Total
1966	74.70	78.0
1970	135.0	138.7
1971	134.72	138.26
1972	133.83	137.07
1973	133.68	136.83
1974	136.85	141.04
1975	141.25	144.58
1976	110.10*	151.4

* Proved
Source: MEES 21.2.77 (FT 21.3.77)

Although Saudi Arabia had participated in the symbolic oil embargo which took place as a result of the 1967 June War, this was

46 A HOUSE BUILT ON SAND

more through King Faisal's concern to gain Muslim access to the Holy Places of Palestine and it was only under considerable pressure from more militant states that Saudi Arabia participated in the struggle to improve its position relative to the companies, after the Teheran agreement in 1971. There have been three main phases in this development: firstly the desire to increase 'participation', secondly the 1973 October War and the oil embargo which followed it, making the first overtly political use of Saudi oil, and thirdly the reversion to the policy of keeping oil out of Middle East politics, and the inconclusive debate on prices since 1974.

The concept of 'participation' was developed in the late 1960s by Sheikh Yamani to counter the growing demands in the Arab world for outright nationalisation of oil resources. This pressure came from progressive and nationalist governments like Libya, Algeria and Iraq, who were gradually working to achieve complete control of their oil resources. On the whole 'participation' was not clearly defined, but mainly dicussed in terms of being a form of non-nationalisation. The reason for this is made clear by Joe Stork, who does give a clear definition:

> 'Nationalisation refers to the takeover by a nation state of the assets and resources of a private individual or corporate entity. In the case of Middle East oil these private entities have been, without exception, foreign owned... Participation means, on the face of it, a state company partnership in which ownership, control and rewards (profits) are shared proportionately. Unless the share is specified this is a rather meaningless term. One hundred per cent participation is, by our definition, nationalisation. Fifty-one per cent nationalisation is, by our definition, participation;... majority participation, or even nationalisation, acquires significance with exercise of ownership and control that is real and not nominal.'[10]

Taking the lead of the reactionary Arab oil producers, Sheikh Yamani started negotiations with the oil companies to achieve 'participation' in 1972. The western press cooperated with Yamani's attempt to set himself up as a nationalist and published lengthy articles and interviews with him, giving the appearance of a confrontation situation between the oil companies and the

---

10. Joe Stork, *op. cit.* p.181-2.

OIL 47

negotiating team. Negotiating on behalf of the Gulf states, Yamani reached an agreement in New York in October 1972. This was hailed as a satisfactory deal for all sides: it gave the producing states an initial 25 per cent ownership immediately, rising to 30 per cent in 1979, and thereafter by 5 per cent yearly until the magic figure of 51 per cent was reached in 1983. Giving minute amounts of oil to the producer states' national oil companies, the agreement left control of operations and management in the hands of the oil companies, thus revealing the extent to which 'participation' was in fact tokenism. Contrary to expectation, not all the countries on whose behalf Yamani had negotiated ratified the agreement: in Kuwait the National Assembly, which was noted for its independence from the Ruling Family, refused, as it considered that its terms did not represent a structural change in the relationship between producer state and oil company.

Further developments in Saudi-style participation were interrupted by the October War of 1973 and the oil embargo, which totally transformed the context of oil negotiations. By the summer of 1974, following on the actions of other oil producers in the Gulf, Saudi Arabia announced its takeover of 60 per cent of Aramco's assets. Since that time desultory negotiations have been going on for complete takeover of Aramco by the government, with no concrete result, although by the end of 1976 all the other Gulf states had gained total ownership of their oil facilities.

Saudi Arabia's 100 per cent takeover of Aramco raises some interesting questions. Although the Saudi Arabian government claims to have taken over 60 per cent of the company's assets in 1974, it only paid for the initial 25 per cent participation agreement.[11] Further:

> 'the US oil companies have until now remained 100 per cent owners of Aramco itself, the Saudi government for more than two years has held 60 per cent of the producing assets of the company'.[12]

Confirming the fact that changes in shareholding do not necessarily mean changes in control of a company, the *International Herald Tribune* adds:

> 'As Aramco likes to point out, the company has already passed

11. *Financial Times*, London, 25.11.76.
12. *Wall Street Journal*, New York, 15.11.76.

48 A HOUSE BUILT ON SAND

through phases of 25, 40 and then 60 per cent government ownership with no other than an accounting effect on its operations. Neither the 40 nor the 60 per cent Saudi ownership has ever been put into a formal document.'[13]

Similarly, throughout the period of negotiation, Aramco, far from winding down its operations, has been expanding and developing its activities, hiring new staff from all over the planet, setting up higher and higher budgets, and taking on massive oil-related projects on behalf of the Saudi Arabian government, such as the mass gas gathering plant and the integrated power system for the eastern region.

Despite these signs of Aramco permanence, the newspaper reading public has seen a series of headlines of the variety 'Aramco-Saudi agreement completed', followed shortly by 'Aramco-Saudi talks postponed', etc. Talks have taken place with greater or lesser secrecy in London, Geneva, Quiberon and elsewhere. On one occasion the meeting caused a considerable stir, when Aramco and Saudi Arabian representatives met in Florida, in supposed secret, under the protection of '50 shotgun-carrying uniformed and plainclothed Wackenhurst security guards, as well as local police, US marshalls, Secret Service agents and US customs agents'.[14] As a public relations officer pointed out, 'how did the participants think they could go down to Florida with their private planes, their helicopters and their shotgun-toting guards and hope to keep a meeting like this secret?'[15]

By November 1976, the agreement was said to include the following:

— compensation on the present book value of $2,000 million, $500 million of which has already been paid (in 1974)...

— separate agreements for further exploration in which companies will lose capital on unsuccessful exploration, but will be refunded when commercial quantities of oil are discovered.

— continued access to the oil and the marketing of the oil by Aramco constituent companies.

— the company to receive a fee of between 10 and 13 cents per barrel for operational and other services.

Although they are slightly different from terms mentioned on

13. *International Herald Tribune*, Paris, 17.6.75.
14. *Washington Post*, Washington, D.C., 8.3.76.
15. *Wall Street Journal*, New York, 15.3.76.

earlier occasions, these conditions are believed to be agreable to all parties who are merely continuing to discuss 'unresolved details'. In the first half of 1977, speculation continued, with little information reaching the public, but the *Financial Times* announced in April that the statutes for the new agreement had been completed. It is generally believed that the agreement will be backdated to apply to 1 January 1976, thus limiting the significance of the delay in relation to the final settlement.

Whatever the cause for the delay, it is clear that serious disagreements exist between the companies, who do not want to lose their golden egg, and the Saudi Arabians as, by late 1976, all other Gulf states had 100 per cent ownership of their oil resources. It also seems that some debate is going on within the Saudi Arabian ruling group on the issue of the settlement, between those who favour the continuation of a major role for Aramco in production and marketing decisions, and those who would prefer far greater control of the entire oil industry in Saudi Arabia by Saudi Arabians, with far less dependence on the Aramco companies. It is true that nationalisation need not give the owning state control of the industry if the state is dependent on a single buyer, which was the case in Mossadegh's Iran, and the cause of his downfall. But the situation is different today when until recently consumers have been prepared to cut each other's throats for a few barrels of oil. Saudi Arabia could easily sell more of its oil to other companies or governments than it has been doing so far. One level at which Saudi Arabia has shown foresight has been in insisting that Aramco train Saudi Arabians for the higher positions in the company at all levels, giving them greater potential for independence from the oil monopolies: in 1976 Aramco employed 20,067 people in Saudi Arabia, 14,837 of whom were Saudi Arabians and only 1,722 Americans; Saudi Arabians held 46 per cent of Aramco's 1,091 supervisory positions. The regime therefore now has the possibility of controlling its own oil industry if it so chooses.

Another important question which is under debate relates to the amount of oil Saudi Arabia should extract. Currently the government's income from oil is well above its needs to sustain its development plans and its current expenditure. Given the scarcity of world resources and lack of value of paper assets, it would be far more beneficial in the long run for the government to keep its oil in the ground and extract it at a rate compatible with its development needs, i.e. about 5 million barrels a day, rather than the 8 to 9 currently being pumped. Such a policy would ensure continued resources for the country for a far longer period of time. Although there are arguments that oil will soon be useless and therefore should be extracted now while it is in demand, these ignore the complexities and problems arising in the search for alternative energy resources. Besides, even if sun or nuclear power were to take over from oil as energy, oil will continue as an essential raw material for hydrocarbon-based industries, which are of major importance in the world economy, and consequently Saudi Arabia could not fail to win by keeping its oil in the ground and extracting it only to finance its internal requirements. As well as the above purely capitalist argument in favour of restriction of oil production, there is a conservationist one.

According to current estimates of reserves (see table p.45 above), under controlled increased production, present reserves would last up to 50 years. As we have noted, oil has many uses and undoubtedly will continue to be a particularly valuable resource. To continue to extract it in such a way as to ensure that in 50 years (at most) there will be none left in Saudi Arabia (therefore probably in the world, as Saudi Arabia has the largest reserves), reveals the total lack of consideration currently given to future generations, whose survival can be put in doubt by the contemporary effort to deprive them of any natural resources.

Although the regime claims to be planning and preparing for an oil-less future, its current oil policies give no indication of such intentions. On the contrary, Saudi Arabia's oil policies appear to be determined by considerations of immediate international politics, rather than the long-term interests of the inhabitants of Saudi Arabia, who will have little use for grandiose petrochemical plants when there is no oil to fuel them.

The immediate political objectives of the Saudi regime's oil policies are only marginally related to the internal situation in the country. They are primarily concerned with the ambitions of the regime as a major international power, both in the Middle East and on the world scene. This is clear from the regime's actions during and after the October War, the only occasion on which it admitted to using oil as a 'weapon', and in its subsequent attempt to restrain oil prices within OPEC.

The October 1973 Middle East war not only gave OPEC the

opportunity at last unilaterally to determine the price of oil, but it also marked the only occasion on which Saudi Arabia actually supported and implemented an oil embargo. After 1967 King Faisal rapidly established Saudi Arabia's supremacy in Arab politics: his ability to offer poor states large subsidies to help with the war effort rapidly allowed the regime to be seen as the leader of the Arab camp in the Arab-Israeli conflict. After Nasser's death in 1970, Faisal's task was made easier by President Sadat's pro-western positions. Having encouraged Sadat to rely on the USA, rather than the USSR, for a solution to his problems with Israel, King Faisal found that by mid-1973 his own efforts to encourage American intervention on the Arab side had failed to bring Israel any closer to compromise.

It was in this context that for the first time there was mention of the possibility of using oil as a weapon in the Middle East problem. Before 1973 Saudi Arabian officials had strenuously denied any such intention whenever it was mentioned by any other Arabs. In the summer months of 1973, Saudi Arabian high officials started making noises indicating that in the event of another Arab-Israeli war they would be unable to stand idly by, and would use the 'oil weapon'. Even once the war had started the regime was reluctant to declare an embargo, and tried to postpone meetings. However, once President Nixon started to re-supply Israel by emergency airlift with all its needs, it was no longer possible for Saudi Arabia to stay inactive. Arab oil producers met and declared an embargo on exports to pro-Israeli states and a reduction in production which would increase by 5 per cent each month during which no progress was made towards restitution of Arab lands.

In early 1974 the embargo was gradually lifted (despite the total absence of progress on the Arab-Israeli conflict) and western states started to come to grips with the new oil price situation, as prices had risen by 470 per cent over the previous year. The solutions envisaged included on the one hand carefully leaked plans for the invasion and occupation of unspecified desert oil fields, and on the other the creation of the concept of 'recycling' petrodollars. This euphemism means ensuring that whatever amounts the west spends on Arab oil is recovered as fast as possible by banking the money in the west or by selling various more or less useless goods and services to oil-rich Arabs at highly inflated prices. In this way the arms trade to oil-rich states has risen dramatically including both obsolete and unsuitable technology and highly advanced weapons which the buyers are not expected to be able to use competently.

52 A HOUSE BUILT ON SAND

These policies were accompanied in the west by a massive campaign portraying Arab oil producers as a small group of blackmailers who were already so rich they didn't know what to do with their money, and moreover, whose policies hurt the poor consumers who were hit by inflation worst of all. The fact that oil company profits were multiplied was glossed over; centuries of western exploitation of the Arab world, and decades of oil companies' outrageous profits at the expense of local populations in producer states were ignored. The Arabs were portrayed in the west as being totally responsible for all the problems of the decay of capitalism, while they drove Cadillacs encrusted with diamonds in the desert. The poverty of the Arab masses was excluded from the picture.

This image persisted despite the fact that one of the staunchest supporters of western capitalism was then, and remains, Saudi Arabia. Since the late 1973 oil price rises, Saudi Arabia's oil policies have clearly supported western capitalism. Saudi Arabia has consistently refused further oil prices rises within OPEC, using the pretext that rises cause inflation in the west, while ignoring the western-sponsored inflation which has ravaged most of the oil-producing states. Saudi Arabia's concern for western economies has led it to spend vast amounts of money propping up European regimes which the ruling Saudi family feared might fall to communism (enemy number one of the Saudis), such as Italy, France, and Spain where contracts have been awarded to support anti-left wing regimes.

While other OPEC countries called for price rises which would at least cover the inflation they had imported from the west, Saudi Arabia opposed this, and in December 1976, at the Doha OPEC meeting where the oil producers failed to agree on oil prices, the organisation's unity was finally broken. Saudi Arabia and the United Arab Emirates increased their prices by 5 per cent while other OPEC countries increased theirs by 10 per cent. At the time the Saudi regime attempted to give the impression that its 'concession' to the West was conditional firstly on an improvement of the West's offer in the North-South dialogue (an unsuccessful conference at which Third World countries attempted to improve their position relative to the capitalist world), and secondly on the USA taking active steps to bring about a Middle East settlement. Developments in the following months clearly revealed that Saudi Arabia was not getting results on either front, although it had created a serious crack in OPEC. Prices were once again unified in

July 1977 when Saudi Arabia and the United Arab Emirates increased their prices by 5 per cent to bring them in line with other OPEC states, but the Saudi Arabian policy had been successful in protecting capitalism as it prevented a much higher price rise and showed that the organisation was no longer a united front.

OPEC had for a short while been a powerful organisation and succeeded in reversing previously highly unfavourable terms of trade. Saudi Arabia's action weakened it by undermining its unity and indicating that it was prepared to go it alone. Since Saudi Arabia produces almost one third of all OPEC oil, its position cannot be ignored.

The present Saudi regime has not only allied itself so closely to the West as to keep oil prices down to benefit western countries, but is even prepared to go so far as to produce its oil even faster. In December 1976 when the split in prices occurred, the government stated its willingness to increase production to meet the expected increased demand for cheaper oil and, indeed, for a few weeks production rose beyond the previously enforced ceiling of 8.5 million barrels a day to close on 10. This move, clearly meant to undermine other OPEC countries, caused considerable hostility, but it shows clearly where the regime stands: alongside western capitalist states and in support of a policy which discourages the world from seeking alternative sources of energy.

Two points stand out from this discussion: first that the Saudi regime's reiterated claim of never interfering in another country's policies is a complete fabrication, as one of the regime's greatest concerns is to influence as many states as possible by giving aid to regimes which are anti-left and pro-Islamic. The second point is that the regime's alliance with the west is, in its own view, essential to its survival.

Petromin and oil related industry

The General Petroleum and Mineral Organisation, better known as Petromin, was founded in November 1962. The government of Saudi Arabia established the organisation because:

> 'The Ministry of Petroleum and Mineral Resources could not, because of its governmental capacity, function as a business organisation in joint ventures involving foreign

investment'.[16]

The creation of this organisation made it possible for an independent company to enter into projects with foreign firms on a straightforward commercial basis. In this sense it is somewhat different from other national oil companies established by producer states as they gained some control over the industry.

Petromin is responsible for managing the oil which accrues to the state as a result of agreements with the concession companies, as all these agreements specify that a certain amount of oil is to be handed over to the state oil company. It is also responsible for prospecting further mineral resources through geological surveys, and for a number of joint projects both with Saudi Arabian private capital and with foreign private and state companies. The Organisation has acquired for the government a number of concessions for the search and exploitation of oil; the first was in 1967 and Petromin assigned it to the Italian company Agip; the second in 1968 was assigned to Phillips. These were located in the Rub'al Khali and Hofuf regions respectively. A third concession covering the Red Sea coast given to Petromin was assigned to a group of companies including the Natomas Internal Corporation, the Sinclair Arabian Oil Company and the Pakistani government. As we have mentioned, Petromin has an option on equity participation should oil be discovered in the Auxirap zone, which has since been redistributed among other companies. Since January 1974 Petromin has been in charge of the Saudi Arabian government's 60 per cent share in Aramco.

In December 1975, after the creation of the Ministry of Industry and Electricity, some of the functions of Petromin were transferred to the Ministry of Industry and its autonomous organisation the Saudi Arabian Basic Industries Corporation. Petromin retains control of the marketing, refining, distribution and transportation of oil, leaving other industrial offshoots to the Ministry of Industry.

As part of its survey operations, Petromin set up two joint companies: *the Arabian Drilling Company (ADC)*, formed in 1964, 51 per cent owned by Petromin, and the rest by two French companies: Forex and Languedocienne Forenco. Technical

---

16. Ahmed Zaki Yamani, in Petromin *Progress Report 1968*, p.5; quoted in R. Knauerhase *The Saudi Arabian Economy*, Praeger Books 1975.

management is by the French and administration by Petromin. The other company is the *Arabian Geophysical and Surveying Company* (*ARGAS*) formed in 1966 with a capital of SR2 million, again 51 per cent owned by Petromin and the rest by the French Compagnie Générale de Géophysique.

Petromin has built two refineries: the *Jeddah Oil Refinery Company*, of which it owns 75 per cent, the rest is owned by the *Saudi Arabian Refining Company* (*SARCO*), which was created in 1967 with a capital of SR150 million; it started production in August 1968 with a capacity of 12,000 barrels per day. This was recently increased to 45,000 barrels per day, in order to enable it to supply the western region with its oil needs. The second refinery, which is in Riyadh, has a capacity of 15,000 barrels per day. Both are scheduled for considerable expansion.

Since 1963 Petromin has been responsible, through its *Marketing Department* for the distribution of oil products in the Kingdom; its sales became significant only after 1967 when an agreement was reached whereby Aramco handed over its distribution rights in the eastern, central and northern regions of the country. At the international level through its *International Trade Division* it is responsible for the sale of royalty oil, i.e. the oil which Aramco hands over as part of its concessionary agreement as a 'royalty' to the producing state.

Another company operating in oil related industry is the *Petromin Lubricating Oil Company* (*Petrolube*), founded in 1967 with 71 per cent ownership by Petromin and the remaining 29 per cent by Mobil and other private interests. It produces basic lubricants in a factory in Jeddah which opened in 1971.

In the transport field, Petromin has set up two companies: the *Petromin Tankers and Mineral Shipping Company* (*Petroship*) was founded in 1968 with a capital of SR10 million; its activities include operation of sea going transport and acquisition of a fleet. It has a number of ships which mainly carry crude from the eastern region to other refineries. The *Saudi Marine Petroleum Construction Company* (*Marinco*) founded in 1968, is 51 per cent owned by Petromin and 49 per cent by the US-owned company McDermott and is involved in sea construction, such as drilling platforms, jetties and the laying of underwater pipelines.

In an attempt to set up some petrochemical industries, Petromin formed the *Saudi Arabian Fertilizer Company* (*SAFCO*) in 1964 with a capital of SR100 million. Since early 1976 SAFCO has become the responsibility of the Saudi Arabian Basic Industries Corporation (SABIC) who owns 51 per cent of the shares, while the Saudi Arabian public owns the rest. The factory was built in al Hasa to use the natural gas which previously was flared in the oil fields. Production started in 1968. Under the management of Occidental Petroleum the factory only produced at 23 per cent of capacity and in 1973 management was taken over by Petromin who worked to increase output to 80 per cent of capacity. Designed to produce 1,100 tonnes a day of urea from 600 tonnes a day of ammonia, the company's maximum production up to 1977 has been 600 tonnes. This is partly due to faulty construction involving the use of brackish water in the cooling system causing corrosion. As a result of such carelessness SAFCO and Occidental are engaged in litigation over the minor profits which were made in 1974 and 1975, the only two years in which there were any profits. The fact that the American company has no share in ownership is now regretted by the Saudi Government who favour maximum joint responsibility in order to avoid similar problems.

These limited offshoots of the oil industry indicate the absence of serious attempts to diversify away from crude oil production in the past. As we will see later there are massive projects for oil related industry in the Second Development Plan, but their value is also questionable.

CHAPTER IV
THE CONTEMPORARY STATE

I Politics and The Ruling Family

Although the borders of Saudi Arabia were defined by 1934, there was little internal political development between that date and the death of Abdel Aziz in 1953: in his lifetime he was unable to transform the country into a nation. The original division between the Hejaz and Najd persisted, and the development of al Hasa as the oil region intensified the differentiation of the eastern region from the rest of the country. By the time of Saud's accession to the throne in 1953 Saudi Arabia was composed of three regions whose conditions and development were totally different. The geographic isolation of these regions from one another increased their people's tendency to consider themselves as separate groups.

In the early days of the Saudi regime, the family had no real control over the country. It was merely the strongest among a fairly large group of tribal leaders, each of whom controlled his own tribe independently of others. Abdel Aziz had succeeded in transforming this situation of near equality among leaders into one in which he had total overall and hereditary control. To achieve this he formed strong alliances with certain tribal leaders, the most important being the Sudairi and the al-Shaykh (the direct descendants of Abdel Wahhab). He made a point of marrying into the tribes he had conquered, a policy which forged unity on the one hand, but created conditions for disunity within the royal family on the other, derived from tensions between members with different maternal tribal affiliations. His relatives and allies were given very powerful positions. Many of them are today regional governors, such as the Sudairi in Asir and the ibn Jiluwi, whose control of the eastern region appears to be hereditary and largely autonomous. Local government is still, within certain limits, somewhat independent from central government.

During Abdel Aziz's reign, while the traditional *majlis* continued to meet, the King adopted a strategy of relying on personal advisors, most of whom were foreign, as a means of establishing his independence from the tribal leaders. These advisors came from Syria, Egypt, and other Arab countries, settled in Saudi Arabia and derived considerable benefits from their position. Through their dependency they were totally devoted to the King and his regime, and since they were educated, they brought an understanding of Arab and world affairs to the regime, unlike the isolated Wahhabis who had little knowledge of the outside world.

Subsidies to tribal leaders to retain their loyalty remained a traditional means of political control in the country, and Abdel Aziz was able to fulfil this obligation thanks to the increasing revenue from the oil concession. Although peanuts compared to the income accruing to the oil companies, by 1950 the regime was receiving $56 million a year in oil revenue, a bonanza which allowed the King to distribute greater sums than ever to tribal leaders. Since this new income came from a source totally outside the traditional economy and accrued directly to the royal family, the latter's power was further increased and other forces weakened.

Through this combination of traditional methods of control and the distribution of the new-found wealth, Abdel Aziz was able to strengthen his power relative to other leaders and to become in fact as well as in name a King, although his kingdom was not really unified, remaining a collection of disparate tribal groups, ruled by their local leaders, some of whom had influence in the royal palace. Hardly any attempts were made to build new political structures or to use the oil revenues to develop the country. Apart from the subsidies to tribal leaders, the income was used for items of conspicuous consumption for the ruling family — palaces, cars, water pumps and generators to provide electricity and water for the palace gardens, and airports in Riyadh and Jeddah. The Jeddah port was enlarged in the late 1940s as imports increased and a few tarmaced roads built between the main centres for the royal vehicles.

Finances continued to be in the hands of Sulaiman al Hamdan, whose main pre-occupation was to enrich himself and the royal family. There was no financial management of any kind and spending was uncontrolled. Under pressure from the United States, Abdel Aziz finally allowed the creation of the Saudi Arabian Monetary Agency (SAMA) in 1952 under the direction of an American. It had little effect. In the last years of his life Abdel Aziz realised the necessity of establishing some sort of state structure and in the last days of his reign he appointed a Council of Ministers. The Council never met during Abdel Aziz's lifetime.

After Abdel Aziz's death in 1953, King Saud continued his father's tradition of personal rule with the assistance of foreign advisors, only slightly tempered by the new Council of Ministers. The Council was chaired by Saud and his brother Faisal, who was also Minister of Foreign Affairs, was vice-chairman, as well as Crown Prince. The Council had a merely advisory role and included both Ministers and advisors, all appointed by the King.

Saud ruled in even more extravagant style than his father. A budget, the first ever in the kingdom, was published in 1954, but remained a dead letter, and the performance was not repeated for some years. The regime's fanciful approach to finances continued, with spending regulated according to the King's whim. A number of new palaces were built at great cost — Nasiriyah in Riyadh ($25 million), Jeddah ($25 million) — and the King spent a great deal of time travelling through the country to ensure continued support from the tribes, lavishing their leaders with gifts. The subsidy of tribal leaders continued to be a basic element in keeping political support for the regime, and was seen as a necessary expense. This distribution of the country's resources to the favoured few laid the basis for a future privileged class who learned to use their positions in government for private gain.

While the rapid increase in oil revenues financed a rash of wild consumerism, it was also used to build the bare bones of an infrastructure. Any expenditure in this direction mainly benefitted

1. For an account of life in King Saud's palaces, read J. Arnold, *Golden Swords and Pots and Pans,* London, 1964. This is a hilarious description of palace politics and corruption by a chief steward, whose perception of events is very acute.

the US building companies and exporters involved. Saud did, however, use some of the now comparatively massive revenues on such socially necessary projects as schools and hospitals, although not nearly enough to meet the needs of the ordinary people. These minimal projects were staffed by expatriates from other Arab countries, and gave the people their first benefits from oil revenues.

The considerable Egyptian presence in the country, including teachers and military training missions, as well as other professions played a role in arousing the interest of the local population in Arab nationalism. The ideas of Nasserism had already penetrated the country's isolation through Cairo's 'Voice of the Arabs' radio broadcasts, which were avidly listened to. A few Saudi Arabian boys had been going to Egypt for secondary education since the 1940s; some of them were influenced by Arab nationalism. The impact of nationalism was regarded with thorough distaste by the still powerful religious authorities, who also disapproved of the ruling minority's dissolute way of life. In order to appease them the regime had to take some action, and in 1955 issued a decree forbidding Saudi Arabian boys from going to schools abroad. This was justified by the argument that there were enough schools within the kingdom. Regardless of the truth of this, there certainly weren't enough teachers, so the effect of the decree was to increase the numbers imported from Egypt, thus strengthening, rather than diminishing, the nationalist influence.

Other concessions to the religious leaders included the strengthening of the Public Morality Committees, or 'religious police' who were empowered to enter people's homes to force them to pray at the right times of day, and to see to it that no vices, such as smoking, were indulged.

The 1958 crisis

Although oil revenues had risen from $56 million in 1950 to $341 million in 1955, the country was in the throes of a financial crisis by 1956. The crisis was brought about by a combination of factors: to the total absence of budgeting and uncontrolled government spending based on royal spontaneity, was added the withdrawal of credit facilities. Aramco was reducing its production in Saudi Arabia, in accordance with its world marketing plans, and refused advances on royalties, claiming that the price of oil was falling. The company discouraged other international bodies from replacing them in loaning cash to the regime, which was not only demanding more money but also calling on the company to ensure the completion of extravagant projects. Confidence in the regime collapsed as contractors remained unpaid for completed projects while being asked to start work on new and equally costly ones, importers were not paid for goods delivered, and government employees were paid erratically, if at all. Gold and foreign currency reserves fell to their lowest ever, the rial's market value sunk to about half its official value, and the public debt rose by late 1957 to 1,800 million rials.

Added to the financial crisis was a political one. Suez not only contributed to the decline in oil production and revenues, it also gave a great boost to Arab nationalism, then in its most active phase. Nasserism had reached far beyond the borders of Egypt and nationalist positions gained widespread support amongst educated people who opposed western supremacy in the Arab world and detested western support for Israel.

At the internal level, Saud wavered between support for the religious and extremist tendencies in his regime which favoured the continuing relationship with the USA, and on the other hand support for the more progressive Arab nationalist elements, both in the country and more generally in the Arab world. In July 1956 the oil workers, the first elements of a working class in Saudi Arabia, went on strike for a combination of economic and nationalist demands, presenting a danger feared by all sections of the ruling group. Saud's confusion between an ideological leaning towards nationalism and his need for continued US support for his extravagance had led him to attempt to lead the Arab world. After Suez, he did this by supporting Eisenhower who pronounced that 'the existing vacuum in the Middle East must be filled by the US before it is filled by the Russians'.[2] This finalised the breach with the nationalists, which had begun with Saud's response to the strike, and formalised the division of the Arab world between the nationalists and pro-imperialists.

Saud's indecision, added to what had become total financial chaos, led to his downfall. In early 1958 it was disclosed that Saud had financially sponsored an assassination attempt on Nasser. This revelation caused an outcry against the regime, both internally and in the Arab world as a whole, and also rendered Saud's personal intervention useless to US strategy in the Middle East. The

2. Statement quoted in F. Halliday, *Arabia without Sultans*, London 1974, p.64.

Council of Senior Princes, responding to the general dissatisfaction among the more influential sectors of the country, and to the promptings from Washington, deemed it advisable to hand over full executive power to Prince Faisal, who was named Prime Minister, and instructed to solve the political and financial crisis.

Faisal's first step, on gaining control, was to introduce a programme of financial reforms under the advice of the International Monetary Fund (IMF) and the International Bank of Reconstruction and Development (the World Bank). The state became a member of the IMF and the World Bank and the measures decided upon included cuts in the allowances of the royal household. As a result some dissastisfaction developed within parts of the royal family, and among the group of importers who had benefitted from the trade in luxury goods. These factors helped Saud to return to power in 1960, but the 1958 austerity budget was a success bringing the country's finances to solvency within 18 months. By January 1960 gold and foreign currency reserves had risen from $24 million to $186 million, prices had declined by 15-20 per cent and the government had paid off debts of $180,733,000 to SAMA. In 1960 the rial was officially devalued to SR4.50 against the dollar, and by 1961 the vast deficit was transformed into a surplus.

No solution was found, however, to the political contradictions which gripped the regime. There was pressure from the first generation of educated people who wanted to see the country developed, and there was opposite pressure from the religious leaders who were horrified at the dissolute life-style of the royal entourage and at the western influence which was penetrating the country in its most obnoxious forms. There was conflict between supporters of western imperialism and Arab nationalism, which was sweeping across the Middle East. It was in this context that the different elements of the ruling family and its entourage struggled for power between 1958 and 1962, a struggle between the Faisal and Saud factions. As Saud had previously been defeated by economically stabilising policies combined with political conservatism, he changed his position and formed an alliance with the liberal faction within the family, which later came to be known as the 'Free Princes'. Saud was seen successively as a wasteful dissolute spender and a supporter of democratic reforms and of nationalism. Faisal, by contrast, was seen as a competent negotiator who opposed political reform but supported economic modernisation; he was also known to have the support of the USA, which had come to distrust Saud's fickleness.

In 1960 Saud found himself strong enough to oust Faisal and set up his own government in which he was Prime Minister and for the first time some 'progressive' men were given ministries: they included Prince Talal ibn Abdel Aziz, the leader of the 'Free Princes', and a commoner, Abdullah Tariqi (see Chapter 3 above) who was put in charge of the new Ministry of Petroleum. Saud succeeded in wresting power from Faisal because he had mobilised support on a wide front, ranging from liberal princes to the traditional tribal leaders who had been heavily entertained, and from the importers and more spendthrift members of the royal household whom Faisal had alienated by his austerity policy.

Once back in power, Saud soon forgot about his 'progressive' tendencies. Within a few days of the formation of his Cabinet, in late December 1960, it was announced on Mecca Radio that a new constitution had been agreed. This constitution stated that Saudi Arabia was an independent Islamic state and that it would be a constitutional monarchy. There would be a Parliament with a third of its members nominated and the other two-thirds elected. At the economic level national resources were to be owned by the state and there would be cooperation between the private and state sectors of the economy. Trade unions would be allowed to operate. At the social level people would have equal rights and have access to housing, education and health care.

This constitution was never officially proclaimed. In the days following the original announcement, a struggle took place between the liberal princes who had supported it and the conservative forces within the regime. Saud refused to sign the constitution. As a result of this defeat, the militancy of the liberal princes intensified and their criticism of the regime became more virulent. In September 1961, Prince Talal was dismissed from the cabinet, leaving Abdullah Tariqi as the only forward looking progressive in the government. In 1962 Talal went into exile in Cairo where, with three of his brothers, he founded a Front for the Liberation of Arabia.

Having disposed of the progressives, Saud was gradually making the compromises demanded by the rest of the ruling family, but his unreliability and state of health caused concern, and the struggle between the two factions continued well into 1962 when the outbreak of the the Yemeni Civil War once again brought the conflict to a head.

On 26 September 1962 the Imam of Yemen was overthrown and a republican regime proclaimed by the officers who had carried out the coup. Within a week the Imam and his supporters were regrouping on the Saudi border to plan the return of the Imamate, and Egyptian troops had arrived in Sana'a, the capital of Yemen, to assist the new republican regime. The struggle between the republicans, supported by Egypt and the Soviet Union on the one side and the Royalists supported by Saudi Arabia, Britain and the USA on the other, continued until 1967 when, as a result of the June war, Egyptian troops were withdrawn, but the Republicans had made a great enough impact to ensure that the Imam would never return.

King Faisal's reign

By the end of October 1962 King Saud had failed to assist the Imam decisively, and again appeared to be wavering. The Yemeni revolution looked like winning and the now exiled 'Free Princes' had announced the formation of a government in exile. To crown all this, President Nasser made a speech in which he stated that Arab socialism would destroy the Saudi monarchy. The regime was in a state of panic and King Saud quite unable to cope. The Council of Senior Princes once again intervened and gave full powers to Faisal who, this time, was to stay in power, although it was only in 1964 that he was proclaimed King and Saud formally deposed. Saud later went into exile and died in Europe in 1969.

The defection of some Saudi Arabian pilots to Egypt at the beginning of the Yemeni Civil War compounded concern at internal security in Saudi Arabia, particularly as one of them declared that: 'There is an organisation in Saudi Arabia which includes free officers and civilians. All of them are waiting an opportunity, which we hope will come very soon. This organisation was formed before the Yemeni revolution.'[3]

Discontent was clearly at a peak and Faisal felt obliged to announce the Ten Point Programme on 6 November, only a week after taking power. The Programme aimed at quietening

---

3. Statement quoted in *The Times*, London 19.10.62.

dissastisfaction and at raising support for the regime by offering some concessions to popular demands. The first point was the promise of a constitution and a consultative council, which would be established within the Islamic perspective of the state:

'The promulgation of a Basic Law for the government of the country, drawn from the Koran and the Tradition of His Prophet and the acts of the Orthodox Caliphs, that will set forth explicitly the fundamental principles of government and the relationship between the governor and the governed, organise the various powers of the State and the relationship among these powers, and provide for the basic rights of the citizen, including the right to freely express his opinion within the limit of Islamic belief and public policy.'[4]

As we have seen this was not the first mention of a constitution: one had been announced under the influence of the 'Free Princes' in 1960. It was not to be the last: after Faisal's assassination in 1975 the new Crown Prince, Fahd, announced that a consultative assembly and a constitution would be set up. Soon after coming to power Faisal got into the habit of replying to any questions on the subject in the following way:

'A constitution, what for? The Koran is the oldest and most efficient constitution in the world. Elections, a Parliament? After the unfortunate experiments which have been attempted in neighbouring countries, it is better to forget all about it. Believe me, Islam is a sufficiently flexible and far-sighted religion to ensure the happiness of our people.'[5]

The other points in the Programme have since been implemented to a varying extent. They were:

1. The establishment of new local government structures with their own regulations. Although announced in 1963, the local government structures are still not being implemented for a variety of reasons (see below for discussion).

2. An independent judiciary would be set up with a Ministry of Justice. This was only established in 1970.

3. A Judicial Council was to be established which included 20 members from the 'Ulema and lay jurists.

4. Islamic Propaganda was to be strengthened.

5. The Public Morality Committees were to be reformed. They

---

4. Quoted in Gerald de Gaury, *Faisal*, London 1966 p.148.
5. Quoted in *Le Monde*, Paris 24.6.66.

had aroused considerable hostility due to the violence with which their members carried out their duties, and it was hoped that their responsibilities would be restricted, but they were not.

6. Education, health and social security legislation would be enacted, including the protection of workers and provision of recreation facilities. (The Social Security Law was announced in 1962 and Labour Regulations in 1969.)

7. Legislation would be enacted to assure economic progress and expansion, and the increase of capital investment. (Many Decrees have been announced on this issue, including the Foreign Investment Code and various attempts to encourage local investment.)

8. The country's resources and economy were to be developed, e.g. roads, water, industry, agriculture.

9. Slavery was to be abolished throughout the Kingdom. This was immediately decreed and applied over the next year, when compensation to be paid by the state to the former slave owners was worked out and paid. For the ex-slaves, however, this made little immediate difference as most remained servants in their former masters' homes. Few had any resources to try and establish any new life for themselves. The abolition of slavery had become a necessity: outside pressure from the USA and elsewhere had finally impressed on the regime that slavery seriously tarnished its public image abroad, as it was the only state (apart from Oman) where slavery was still officially allowed.

The programme was valuable to help Faisal win the support of the young educated class whom he was providing with employment in the enlarged bureaucracy, and who therefore believed they had a greater influence over developments in the country. Similarly, it could be expected that measures would be taken to encourage the development of an educational and health infrastructure, and that the government would be run in a reliable way. He soon obtained the support of the trading community who appreciated his support for modernisation, although he wanted to restrain its speed. This in any case was not a problem until oil income multiplied in the mid-1970s. His personal asceticism and moral conservatism ensured good relations with the religious authorities, despite a number of clashes, such as the opening of television in 1965.

Having quietened the progressives with the carrot of the Reform Programme, Faisal rapidly regained his grip on the situation. With the control over the centres of power and his re-established alliance with the religious groups and the emerging capitalist sector, Faisal took strong measures against internal dissidence. He imprisoned opponents, and actively supported the Imamites in the Yemen War, even coming slightly in conflict with the USA over this issue. Under Kennedy, the USA went through a brief period of supporting democratic moves in third world countries, in order to deny these places to USSR influence. This phase lasted long enough to cause some trouble over Yemen, as the Kennedy regime recognised the YAR (Yemen Arab Republic) but soon found itself giving the Imamate troops air cover from its Saudi Arabian base, i.e. they were attacking a regime which they had just recognised. Faisal soon put enough pressure on the US to ensure its support for the Imamate cause.

Although the Ten Point Programme was hardly an open door to massive social change, the 1960s can be regarded as the decade during which the first serious elements of an infrastructure were set up in Saudi Arabia. This was largely due to King Faisal's determination to modernise the country and provide its people with some of the benefits which the regime could afford thanks to its oil revenues. Expansion of communications of all kinds, greatly needed in a country as dispersed and vast as Saudi Arabia, helped to create a sense of national identity and reduce regionalism. The development of education from 1960 when there were only 113,176 boys at school to 1970 when there were 385,841 is quite significant. As for girls, their numbers at school rose from practically nil in 1960, as no state schools for girls then existed, to 126,230 in 1970.

Faisal tried to gain popular support by spreading the benefits of oil revenue more widely, creating some social services and providing more employment — some in industrial jobs but mostly in the administration. Developments in the various sectors are discussed below.

At the political level, the Yemen war simmered on until 1970 causing occasional difficulties for the Saudi regime. During that period there was considerable underground activity by nationalist groups, supported by the Egyptian regime which publicised their activities, subsidised them and provided them with weapons and ammunition. The regime imposed severe punishment on anyone caught or suspected of political activity. The regime's position, both internally and internationally, was strengthened by the June

War as, for the first time, Saudi Arabia held a dominant position in Arab politics. This again reduced discontent with the internal situation as some at last saw their country taking a leading role on the Palestinian question, and were beginning to feel benefits of the social and economic reforms.

Faisal's reign was marked by the King's personality and policies, since he dominated the political scene and gave less leeway to his Ministers than has existed since. Internally he was totally uncompromising on politics and refused to allow any liberalisation of the political system: it was during his reign that repression of opponents was most violent, and the mention of political reform was taboo. Although he initiated and encouraged some social and economic reforms, assisting the development of health and educational facilities for the ordinary people of his country, he even objected to the concept 'social change' which in 1970 he associated with 'revolutionary change'. As a conservative he sought to moderate the cultural impact on the society of the various structural changes which were being introduced, and he ensured the continued activities of the Public Morality Committees which he considered to be a stabilising force asserting the dominance of traditional values. At the international level, Faisal's desire to pray in Jerusalem (Islam's third holiest place) was tempered by his determination not to come into conflict with the USA and his fanatical anti-communism. But his desire to make Saudi Arabia one of the major powers was implemented by the manipulation of religion as a unifying factor, and the use of Saudi wealth to gain support.

Faisal's assassination and succession

In the early 1970s, the idea developed in the minds of western observers that Faisal was an essential element to ensure stability in Saudi Arabia and the Middle East; his prestige had grown enormously and his success in balancing the demands of rapid modernisers and traditionalist religious forces in his country had proved his diplomatic skills, as well as his firmness with opponents. His internal policies coincided with the interests of the US insofar as conservatism dominated. His international policies were regarded with favour: he worked to unite as many Islamic and third world countries as possible under the banner of Islam and anti-communism and, although his policy on Palestine sounded tough, few people expected him to be militant when the time came.

This proved to be correct when it was seen that Saudi Arabia was reluctant to enforce the 1973 oil embargo and dropped it at the earliest excuse. Although prepared to show its muscle to the world, the Saudi regime proved that it had no intention of threatening US imperialism: on the contrary, the US benefitted from the embargo and price hikes at the expense of European and Japanese capitalism and the embargo was abandoned within a few months without there having been the slightest sign of compromise from the Israelis or the Americans on the Palestinian issue. By the mid-1970s Saudi Arabia's support was considered essential to any 'Middle East solution' and Faisal was seen as the most powerful pro-western ruler in the Arab world. Although 10 years earlier every major article written on Saudi Arabia predicted certain revolution within 10 years, Faisal proved this wrong by establishing a stable regime which was seen to maintain a balance between modernisation (i.e. the recycling of oil revenues to the western capitalist states by spending it on more or less necessary and grandiose projects) and the maintenance of the basic traditions of Wahhabism (religious puritanism and the absence of democracy) in a very conservative state.

In this context, Faisal's assassination on 25 March 1975 came as a shock. Speculation on the motives and sponsors of the assassination was wild and although it was generally agreed that Faisal ibn Musa'id ibn Abdel Aziz had actually killed the King, everyone suspected some great power to be responsible for the action.

The Prince himself, educated in the US, had shown leftist leanings. He was also known to have family reasons for wanting to get rid of Faisal who had been responsible for his brother's death in 1965. In September of that year his brother, Khaled ibn Musa'id had led a group of religious fanatics in an attack on the newly built television station in Jeddah. In the battle which ensued Khaled was killed, and it is interesting to note that he was officially described as 'mad and subject to mystic fits'. But it seems more likely that Faisal's assassination was motivated by the young prince's progressive tendencies.

Although collective action is often seen as the only way to bring about progressive change in most countries, there was little prospect of a mass uprising in the foreseeable future. Perhaps the assassin believed that if Faisal were removed from the throne, the

70 A HOUSE BUILT ON SAND

stability he had created would end, and the different factions in the ruling family would get involved in a long and vicious struggle for power, allowing the different political tendencies in the kingdom to develop; this type of argument is equally valid at any given time, since the internal situation was structurally static.

On the other hand the international Arab situation was changing in March 1975: since the war in October 1973, King Faisal had played a central role in the diplomatic moves aimed at a pro-western settlement of the Palestine issue: he fully supported Sadat and had brought Assad towards a pro-US position. Kissinger had been commuting all over the Middle East in his attempts to produce some signed agreements. By March 1975 it was clear that all his efforts had failed: although he had succeeded in obtaining all the necessary concessions from the Arab states, the Israelis were as intransigent as ever and refused any deal on the Golan Heights or anywhere else, and the Kissinger shuttle had finally come to its ignominious end, just a week before Faisal's assassination. Since Faisal appeared to be the main factor holding the Arabs together in a belief in the American solution, it may have appeared that to assassinate him could finally put an end to this line of action, and leave the road open to the Palestinians to pursue their objectives without interference from the weakened supporters of an American solution. This may well be how Prince Faisal ibn Musa'id saw the situation, and why he timed his action in this way.

This interpretation of the situation, although widespread, proved to be wrong in all its aspects. First, the royal family did not break up into rival factions violently struggling with each other. On the contrary, within an hour of Faisal's death the Council of Senior Princes met and endorsed the proclamation of Crown Prince Khaled as King, and of Prince Fahd as Crown Prince, thus ending speculation that Fahd would become King. Fahd's prominence in Saudi politics had led people to believe that he would supercede Khaled who had been living quietly due to his health problems, and who had limited his political activities to retaining good relations with the tribes. The solution found showed that the dissension expected between the Sudairi, led by Fahd, who appeared dominant, and others led by Khaled and Abdullah, Commander of the National Guard, had not materialised. Internationally it turned out that the chief so-called progressive leader, Assad of Syria, far from waiting for the opportunity to ditch the American plan, continued to support it, and favoured a crack-down on the

THE CONTEMPORARY STATE 71

Palestinian movement, which had once again the strength to try and act independently of the Arab states.

Uncertainty over Faisal's real objectives in Arab politics was revealed by the variety of interpretations put on the motives for Faisal's assassination. The most prevalent theory in the west was that radicals (anyone from the USSR to Palestinian 'terrorists') had planned the assassination to give the progressives the opportunity of gaining power in the region and to deprive it of the Americans' most faithful ally in the Arab world, while creating chaos in OPEC and simultaneously destabilising the most powerful oil producing state. In support of this theory western journalists were busy trying to establish connections between the assassin and the Weathermen in the USA during his stay there, and suggested that he had met 'communists' on various trips abroad. The opposite version was that the American CIA had plotted Faisal's death because of his intransigence over Jerusalem, and in the faith that whoever succeeded him (assumed to be Fahd) would be even more pro-American than his predecessor, but would be less awkard on the Palestinian question.

The possible 'madness' of the assassin, though prominent in the first days after the murder, soon left the headlines, and the explanation that it was an 'individual act', though publicised, was submerged by other speculation. Prince Faisal ibn Musa'id was publicly beheaded in June, after the Saudi authorities had ample time to interrogate him about possible conspiracies. His beheading revealed that the regime did not think he was mad, as according to Islamic law, madmen are not executed.

King Khaled's regime

Despite almost universal prediction to the contrary, Khaled has turned out to be far more than a figurehead and has taken an active interest in Saudi Arabia's politics both internally and internationally.

At home, Khaled has shown his confidence in the stability of the regime by immediately reducing political repression. Only two weeks after acceding to the throne he issued a decree calling for the release of all political prisoners and allowing political exiles to return. This was implemented immediately and although political opponents were not allowed to engage in politics any more than the rest of the population, they were given the opportunity to re-integrate with society.

Fahd who had been expected to rule while Khaled reigned, has found his field of operations circumscribed by Khaled's determination to control developments. The Second Development Plan which was seen as Fahd's brainchild was allowed to continue, but has been somewhat limited in scope as a result of concern about overdependence on the West and the effect that such rapid changes might have on the structure of society. These problems were known to be of concern to Khaled and Abdullah, the leaders of the traditionalist elements within Saudi Arabia. Khaled's long lasting and good relations with the tribal leaders and the religious authorities have had some influence in slowing down the policies which are being implemented and there are people among the 'modernists' who also find the pace of development too fast for local conditions. One of them is the Minister of Planning, Sheikh Hisham Nazer.

Socially, Khaled has encouraged the development of recreational facilities which do not contradict religious puritanism. This is to be seen in the development of sport facilities, including the ambitious projects for establishing football training and pitches in every school, and sending a large team to compete in the Olympics, although there was no likelihood of them winning anything. At the same time there have been no attempts to liberalise social life: drink laws are as strict as ever, and the traditional Islamic judicial code continues to be enforced: men have been beheaded and stoned to death for sexual offences and there is no talk of changing the judicial system, on the contrary, moves to reinstitute Islamic law in other countries have been encouraged.

Politically, once again there were noises of liberalisation and the introduction of various 'consultative' councils immediately after Faisal's death, but no mention has been made of them since.

In foreign policy, King Khaled has brought about a considerable opening up of the country's foreign relations. Far more flexible than his predecessor, he has ignored the pent-up tensions with other Arab states and improved relations with most of them. In the Gulf he visited all the smaller states in early 1976 and, while discussing and solving outstanding problems, he began to put political pressure on them, thus helping to bring about shifts to the right in Bahrain and Kuwait as well as persuading the Emirates to develop more united policies. The People's Democratic Republic of Yemen, whose regime Faisal had been at pains to overthrow, was recognised in March 1976. Since then

Saudi Arabia has been trying to buy off the regime, rather than destroy it militarily.

Similarly, King Khaled's policy on Eritrea has been far more skilful than earlier policies: rather than oppose the liberation movements he has been doing his best to ensure that a centre-right movement is likely to win. In the Lebanese Civil War, King Khaled was the architect of the Riyadh agreement which ended the fighting and installed an Arab peace keeping force in the country in November 1976. This gained him credit both with the Palestinians and with the opposing Lebanese factions.

In conformity with his views of Islam, he has encouraged, in some cases successfully, the return to Islamic law in other Muslim states: in 1977 Egypt, Sudan and Pakistan have all changed their legal codes to comply more closely with the Islamic code.

Overall King Khaled's reign is significantly different from Faisal's. In many ways it is less personal as, at long last, most Ministers are allowed to run their Ministries in their way once major policy decisions have been taken. Power is more widely distributed within the ruling family. The Cabinet announced in October 1975 not only created new ministries which were put under the leadership of specialised technocrats, but also extended the power base within the royal family. To counterbalance the weight of the Sudairi faction which held 3 ministerial posts (First Deputy Premier, Prince Fahd, Defence and Aviation, Prince Sultan, Interior, Prince Nayef), younger members of the royal family were introduced, such as Prince Mutib ibn Abdel Aziz (Public works and Housing) and Prince Majid ibn Abdel Aziz (Municipal and Rural Affairs), and most importantly Prince Saud ibn Faisal, the son of the previous King.

But differences between the two main factions of the family are becoming stronger. On the one hand the Sudairi Seven, all sons of the same mother, have considerable power in the country and are associated with the modernising trend. On the other Prince Abdullah related to the ibn Jiluwi, as are Khaled and his elder brother Mohammed who renounced the throne in 1965, lead the group which supports the maintenance of religious tradition and the integrity of Saudi Arabian society. They find some of the habits introduced as adjuncts of a modern economy distasteful and are concerned at the erosion of traditional values.

II. State Structures

When discussing state structures in Saudi Arabia, it is important to remember that the country has never been occupied by any colonial power, and that until the 20th Century it was a purely tribal society. Therefore it never developed the bureaucratic trappings of empire, as most other third world countries did. At the time of Aramco's arrival, illiteracy was almost 100 per cent and the only 'modern' sector was a small trading community in the Hejaz, an area still largely alienated from the regime because of its liberal culture and its opposition to a puritanical Wahhabism. It is therefore not surprising to find that the present administrative structures suffer from considerable inefficiency, are run by inexperienced people and make few innovations.

Justice

As Saudi Arabia's rulers keep repeating, the Koran is the only constitution of the state, and the best constitution in the world. Based on the Hanbali school of Islamic law, the Saudi regime is the strictest and most conservative of the regimes which still retain Islamic law. Within this structure, the King is also the Imam of Islam, and is responsible for the maintenance of Islamic values throughout the community. Because of the location of Mecca and Medina within the kingdom, the regime also considers itself to be the guardian of Islam and Islamic values throughout the world. Wahhabism being a fundamentalist and proselytising creed, its supporters oppose any liberal tendencies within Islam both at home and abroad.

In Saudi Arabia, the law is interpreted and applied by the 'Ulema, the council of religious sages, all of whom are strict Wahhabis, and consistently oppose modernisation in the kingdom. Their power is still considerable although, as a result of developed relations with the capitalist world, some sections of the Royal family have taken positions conflicting with those of the 'Ulema and forced the latter to concede on certain points, such as the introduction of television. Their influence among the bulk of the population is still considerable: they control the education of girls, and have strong support among the young whose education at all levels has a heavy religious component. Young people tend to adhere to Wahhabi values and decry westernisation, while at the same time they are attracted by it.

The active instruments of the religious authorities are the Public

Morality Committees which are in effect a form of religious police. Their duty is to ensure that the prescriptions of Islam are enforced: that people pray at the correct time, fast when they are supposed to, and do not sin. They are under the control of the al-Shaykh family who hold the main religious positions in the kingdom. They operate like a police force with powers to enter homes to check that religion is properly observed, and may punish those who are found to infringe the rules.

At the administrative level, the first steps were taken when Abdel Aziz created the Council of Ministers, which was made responsible for budgeting. It now has policy-making powers subject to the approval of the King, and legislation takes the form of Royal Decree. There are now twenty ministries, some of which were only set up in the October 1975 reshuffle. At the time of Faisal's accession to the throne, all he had to work with were the skeletons of so-called ministries: Foreign Affairs, which he had established and run since 1930, and Finance which until 1958 had been no more than an institution for handing over funds to the King to spend as he pleased on the one hand, and on the other to keep creditors at bay. A Ministry of Defence was set up in 1946 under the influence of US and UK military missions to integrate the Saudi Armed Forces into a cohesive body. Between 1951 and 1954, posts had been allocated to new Ministers of the Interior, Education, Agriculture, Communications, Commerce and Industry, and Health, but none of these could be properly described as decision making institutions. As noted in Chapter III in 1960 for the first time a Ministry of Petroleum was set up under the leadership of Abdullah Tariqi, the first western-educated expert to hold a position in the government. When he was sacked in 1962 for being too progressive, the position was given to Sheikh Ahmad Zaki Yamani, also educated in the US, who still holds the post.

In 1962 Faisal set up a Ministry for Social and Labour Affairs to implement new regulations, and the Ministry of Pilgrimage as well as that of Information were also created at that time. The Ministry of Justice was created as recently as 1970, although it had been announced since 1962.

The Ministry of Justice is responsible for the application of traditional Shari'a law and formulating legislation for international contracts and other commercial issues which affect the state. The system of criminal justice and of law regarding personal status is still the traditional one, including beheading by sword in a public place for murder or adultery, the amputation of hands for theft and the lapidation of women for adultery. These penalties are still implemented, and defended as being 'good deterrents' by the regime who point to their low crime rate as an endorsement for the policy. Far from considering a change in the legal system, Saudi Arabia is currently trying, and succeeding, to persuade other Islamic states to revert to this judicial code.

In 1954 the Grievance Board was established, as a means to allow citizens to make complaints of a broad nature which did not fall within the scope of Shari'a criteria. It is a committee of the Council of Ministers and can be appealed to on issues which do not concern religious law.

The army and paramilitary forces.

Until the early 1960s the Saudi armed forces were ill-equipped, ill trained and inefficient. Then, as a result of the threat posed both externally and internally by the Yemen Civil War, the regime decided to strengthen its armed forces. In order to retain the balance between the two major sections of the Royal family, the National Guard and the Army have remained two separate bodies with different leadership. This is a unique situation as in most states efforts are made to ensure that the armed forces are under the control of the state, which is usually united enough to be satisfied with one military base, whereas here the differences between the factions within the ruling family are such that two separate forces have been retained.

Since the Second World War there have been a series of foreign training missions in the Saudi Armed Forces. After the British, American and Egyptian mixture in the 1960s, the US acquired almost total monopoly of military training and supplies, after Saudi Arabian fears of Egyptian-led Arab nationalism had brought the regime to end Egyptian training schools, and the dispute between Britain and Saudi Arabia over control of the Buraimi oasis (see below p.112) had led to a break in relations between the two countries.

The Yemen War brought about a massive increase in interest in the Army. As soon as the Yemeni Republic had been proclaimed Britain and Saudi Arabia patched up their differences as both felt threatened by the event; Britain because of the challenge a republic posed for South Arabia. In 1965 a contract for an air defence system was signed involving £120 million for Britain and £20 million for the US. Since then the Saudi Arabian Armed Forces have incurred enormous expenditure on mainly US weapons, and training courses. Despite this it is generally estimated that the Saudi Army and Air Force are unable to cope with the advanced technological hardware with which they are now being equipped and their ability to withstand any military attack is considered to be weak. They have been involved in no action apart from the presence of a mere 750 men sent in late 1976 as part of the so-called peace-keeping force in Lebanon. The weakness of the Saudi forces is shown in that they have in no way intervened in the war in Oman between the Qabus regime and the People's Front for the Liberation of Oman. Although Saudi Arabia was concerned over Iranian military intervention in the Arabian peninsula, the Saudi regime was unable to send its own forces to prevent it.

The Armed Forces currently number 51,500 and there are an estimated 10,000 Americans in Saudi Arabia involved in military-related contracts. Although the Saudi Air Force is equipped with 30 F-5 fighter aircraft (another 100 are on order) the Saudi Arabian government has been 'dissuaded' from purchasing F-14 and F-15 which are extremely advanced technologically and their personnel is hardly likely to be able to cope with its existing equipment within the foreseeable future. Saudi Arabia is also acquiring 1,000 US-made Sidewinder air to air missiles and 650 Maverick air to ground missiles.

Actual defence expenditure quadrupled between 1973 and 1975, jumping from $1,478 million to $6,771 million, representing a per capita expenditure of $1,692 (on the basis of a population estimate of 4 million). There is no sign of any falling off as the massive orders made in the USA in 1975 and 1976 are to be delivered and paid for over the next few years. On top of that, continuing contracts for training and maintenance, such as the renewal of the BAC contract with Britain (£760 million) will escalate defence costs.

Within Saudi Arabia itself the regular Armed Forces are not trusted by the regime. Within their ranks considerable dissent has been known to exist, and it is also notorious that armies have a habit of trying to seize political power in the Arab world; the Air Force did attempt a coup in 1969. Therefore the Army is feared by the regime, and consequently the ruling family is not over-concerned by its obvious inability to deal with any serious threat from outside. The army is however under the leadership of one of the Sudairi brothers, Sultan ibn Abdel Aziz, who stands second in line within the faction led by Prince Fahd.

To counterbalance the threat of the army the Saud have always had recourse to a tribally recruited force. Originally created as the Ikhwan, an army of people who abandoned their tribal allegiance in favour of total devotion to the Saud, this force was destroyed in 1930 when it rebelled against the King on politico-religious grounds. But many of its members joined the White Guard, a paramilitary force which replaced them and which was renamed the National Guard in 1964. At that time the Royal Guard, who were the personal bodyguards of prominent members of the Royal Family had shown their support for King Saud in his struggle with Faisal; they were disbanded and integrated into the regular armed forces. The National Guard fully support the Saud ruling dynasty and are recruited from selected tribes. They are kept isolated from the rest of the population, usually in camps just outside cities, given considerable privileges and encouraged to consider themselves part of the traditional ruling elite. The National Guard is the force trained and equipped to deal with any internal problems, be they tribal revolts, popular unrest, or attempted military coups. In the early 1950s, when there were strikes and demonstrations in the eastern region, units of the National Guard were posted near the cities as a warning to the population.

Formerly equipped with light equipment and trained by the British, the National Guard has recently been reorganised into motorised and fast-moving units. In a $350 million modernisation package, the US are to provide armoured cars and artillery. Training is to be provided by the US-based Vinnell Corporation, who as well as building a training centre are to

> 'Equip and train four mechanised infantry battalions and one light field artillery battalion. The modernisation programme covers training and equipping of approximately 3,600 Saudi Arabian National Guard troops. Their strength consists of 3,000 mechanised infantry troops and 600 field artillery troops. The Saudis will be trained to use V-150 armoured cars, TOW missile system, Vulcan system 81mm

mortar, 105mm Howitzers.'[6]

Currently led by Prince Abdullah ibn Abdel Aziz, the National Guard is composed of about 20,000 men but, despite its small numbers, it is considered to be a more effective force with better training and greater loyalty to the regime. Its rivalry with the armed forces is indicated by Prince Abdullah's insistence recently that the National Guard should be equipped with aircraft, thus giving it access to the same facilities as the Air Force. This has not yet happened, but the National Guard does have helicopters.

Education.

Like most things in Saudi Arabia, education is still in its infancy. Starting with very few resources and mainly foreign (usually Egyptian) teachers in the 1950s, the education system has developed considerably since the 1960s. Curricula, which were traditionally oriented towards religion and under the control of the religious authorities, have been changed to conform more with the Egyptian practice, which is used in most Arab states. More recently modern approaches to pedagogy are being introduced, but only on a small scale.

The Ministry of Education was founded in 1954 and is in charge of boy's education, which expanded rapidly from the 1950s onwards. Until 1960 state girls' schools were forbidden and only a few private schools catering for the daughters of the Hejazi merchant families existed. In 1960, under Faisal's influence, the Directorate General of Girls' Schools was founded, but as a concession to religious opposition to the education of women the religious authorities were put in charge of it, and therefore women's education is under the control of the Directorate and not the Ministry of Education. This also means that girls' education is even more imbued than the rest of the system with the principles of Wahhabism.

In 1952 there were 41,678 pupils at all levels of school education in the state sector. Thereafter the figures grew as follows:

	1952	1960	1965	1970	1973
boys	41,678	113,176,	244,431	385,841	497,733
girls	-	5,224	42,182	126,230	200,786

6. Hearings, Subcommittee on Foreign Assistance, US Senate Committee on Foreign Relations, November and December 1975 p.195.

These figures show considerable progress for 1960 was the first year of state education for girls; they conceal the fact that most students are in the lower levels of education and the numbers in higher education at home are only slowly rising.

There are a number of social problems which are particularly visible in the field of education. The clearest example is the situation prevailing in technical education. From the early 1960s onwards the government has attempted to introduce technical education and by 1976 training centres had a capacity of 2,500 students. These centres are supposed to create a skilled labour force by introducing young men to the technical skills which are increasingly necessary in the country. These centres are on the whole a failure and are never able to find enough students to fill the available places, despite incentives to students in the form of grants, good conditions and teaching. The main reason seems to be sociological: it is a country where, by tradition, some occupations are considered more honourable than others, and where manual work in particular is associated with the less well respected tribes and artisans. Trade and office work are considered honourable, as the former in particular was the activity of the centuries-old Hejazi bourgeoisie and the more noble nomadic tribes. Because of this prejudice, it is difficult to persuade young people whose ambition is to rise in the world, that learning a skill and becoming a motor mechanic, an electrician or a plumber is an acceptable activity for an educated person, even though it can be shown that such skilled craftsmen could make a fair profit. Students and other people, their families in particular, tend to think that anyone with the slightest education should enter the bureaucracy and become automatically a highly respected person. By 1975, only 4,371 people had graduated from the vocational training centres. The most coveted occupations are in government employment or in the import business, at present an easy source of enrichment.

Women's education suffers from other handicaps. There is little incentive for girls to study (except boredom) since their adult career prospects are extremely limited. Except for teaching and medical work in institutions for women only, there are no other recognised openings, whatever level of education has been reached. In the mid-1970s only 1.5 per cent of adult women are actively engaged in work. Particularly at the professional level, this is a comparatively unusual situation in the Arab world, where professional women have been working for decades and form a significant part of the bureaucracy particularly in Egypt.

The universities and higher education institutes have, or are constructing, all the facilities which could be desired, but life is very restricted to formal study and there is little intellectual activity beyond the narrow confines of the educational programme, education being seen as an acquisition of skills rather than knowledge. Women are segregated to their own sections of the universities and kept away from men: lectures from males are by videotape, and special opening hours are arranged for them in libraries.

A further problem in the training and higher education field is the juxtaposition between highly advanced show pieces such as the King Faisal Medical City and the medical schools where existing facilities are particularly inadequate.

Despite considerable efforts to increase the number of Saudi Arabian teachers, the situation is nowhere near self-sufficiency. Expatriate teachers, mainly Palestinian and Egyptian still form 85 per cent of the staff in secondary schools and over half the staff in primary schools, where most of the Saudi Arabian teachers are found. In 1975 there were 4,547 graduates in teacher training for lower levels of education, and 102 for post-secondary education.

Another problem is that students are aware that their likelihood of obtaining satisfactory employment is largely unrelated to their level of educational achievement, or even to the nature of their qualifications, but rather to their family and other personal connections. Hence the standards of educational attainment tend to be low.

Health and welfare

Traditionally regarded as community problems, health and welfare have now been brought within the scope of the administration.

In the health sector, western-trained doctors were brought from abroad to treat members of the royal family since the early part of the century, and the expansion of facilities for the people was slow. In the 1950s there were only about 20 doctors in the entire country, and malaria, trachoma, and enteric diseases were rampant. In the eastern region Aramco started health programmes for its employees and their families from the 1950s onwards, but it was only in the 1960s that the government started spending substantial amounts on social services and health.

At first sight the situation seems to have improved considerably, as the number of hospitals in the country rose from 29 in 1958 to 54 in 1973, and the number of hospital beds from 2,617 to 8,870 in the same period. In the rural areas, the number of clinics rose from 45 to 206 and of health centres from 37 to 360. These figures do not give an exact picture of the situation as on the one hand the so-called health centres are little more than first aid posts and the clinics are staffed by a single doctor, a female nurse and midwife and one or two male nurses.

The most significant problem is, however, the distribution of these facilities. In 1972 of the 51 existing hospitals, 13 were in or around Riyadh, 7 in the eastern region, 12 in the Jeddah, Mecca and Medina area, thus being concentrated in the urban privileged areas and leaving only a very small number for the rural areas such as the north or the relatively highly populated Asir. Again, of the 8,132 hospital beds, 7,115 are in those same areas and in Ta'if which is the summer capital. Even for health centres and dispensaries we find that there are as many in the more developed regions as in the larger and more isolated rural areas.

Here again the regime seems ready to construct large show piece projects which are well publicised abroad and unique on a world scale, such as the King Faisal Specialist Hospital, which is equipped with the most expensive and advanced technology and has therefore to be staffed by expatriate specialists. Such a project is the subject of newspaper supplements and designed to prove to the world that Saudi Arabia is a very advanced country. Meanwhile, ordinary hospitals in the country are waiting for elementary improvements, the local medical students have few facilities, and there are no mobile clinics for the rural and nomadic population.

The 1975-1980 Five Year Development Plan has ambitious objectives for the construction of 70 new district hospitals, mainly located in previously deprived regions, but it remains to be seen how many are built by 1980 and where they will be. Mention is also made in the Plan of mobile clinics, but no details are given.

State-run welfare is a totally new concept for the Saudi Arabians. Traditional tribal structures were based on the extended family as the basic social unit, and it was considered part of the family duty to ensure the welfare of all family members. In this way old people and the sick were cared for and supported by the rest of the group, and this was done without either humiliation on one side, or gratitude on the other, as these things were taken as a matter of course.

In 1962, as part of his programme for social change, Faisal announced a Social Security Law, which was meant to be a first step towards the establishment of comprehensive welfare facilities. The mere need for such an institution marked the breakdown of the traditional structure, and the fact that there were many people in serious need and poverty in the Kingdom. However, given the prevalent Wahhabi thinking and the strong beliefs in self-reliance we find that in one of the wealthiest states of the world, social security is handed out on a means-tested basis to selected groups. The law provides for pensions for orphans, unsupported women, totally disabled people and old people who are totally destitute; if any of these people have another income, the amount of that income is deducted from their pension. The means test, when applied by an already very corrupt administration, is bound to bring about considerable irregularities, particularly benefitting people not in need and excluding genuine cases of people who have no powerful connections.

Further, the social security offices, like most other offices, are located in the richest and most developed areas. They are thus inaccessible to the poorest people in the backward agricultural areas, and the nomads. This point is emphasised by al Awaji:

> 'There are 17 social security centres in Najd compared to 6 in al Hijaz, 6 in al-Hasa and 3 in Asir. Asir contains only 8 per cent of the social security population. Ironically, using the present state of various economic activities in the kingdom, whether industrial or services, we find that Asir has the least of such activity among the four regions. Najd has the central government services and some modern industrial projects. Al-Hijaz on the other hand contains both central government services and industrial activities plus the numerous activities in connection with the annual pilgrimage to Mecca. Al-Hasa includes the oil production which presents its people with many opportunities for earning minimum wages. In contrast, Asir has nothing of the sort. It has only 4.8 per cent of the private establishments in the Kingdom and 3.8 per cent of their labour force. Such facts indicate that the region may have a larger number of unemployed who should benefit

more from the social security system in the country.'[7] The 1969 Social Insurance Regulations introduced a social security system based on contributions from employer and employee to be used for the support of the worker in time of ill-health, or as a pension after the end of his working life. The degree of implementation of these Regulations is unclear. In principle this pension scheme and health care system is reasonably progressive although it puts a very heavy burden of contribution on the workers, in a country where the availability of funds makes it possible for the state to bear the full cost of a comprehensive social security system without endangering other development options.

Community Development Centres were first set up in 1961. As they were a success, another 5 were opened in 1962 and by 1975 there were 17 Community Development Centres, serving 83 communities. Originally set up by the Ministry of Labour and Social Affairs, their objectives are to improve the cultural, social, educational, health and agricultural level of the area in which they are located. This is done by setting up in each centre facilities for literacy classes, public hygiene lessons, libraries, local cooperatives and in the rural areas agricultural advice. Mostly staffed by expatriates, the social development centres have had some impact on rural communities where they have been able to persuade people to come to literacy classes, and to set up producer and consumer cooperatives. Some of these centres, particularly those in areas where nomads have settled, have set up craft workshops, for weaving or other traditional skills, and are in this way keeping these crafts alive and assisting the creation of artisanal industry. In some cases these centres have encouraged men to learn manual skills and become skilled carpenters and other craftsmen. Some of these centres also include health care facilities. Of the 17 existing centres, 11 are located in rural and 6 in urban areas; as they are designed primarily for the non-urban communities, it may be appropriate to ask why there are not more of them in the countryside, and what purpose they serve in the towns.

The media

The regime's ideological control over the country is done through a

---

7. Fatina Amin Shaker, *Modernisation of the developing nations: the case of Saudi Arabia,* unpublished Ph.D, 1972 Purdue University, p.308.

number of means: first and most prominently are the activities of the religious authorities, ranging from the Friday sermons in the mosques to the enforcement of regulations by the Public Morality Committees. The educational system is clearly another method of ideological control. As cinemas are illegal, the media are restricted mainly to the press and television. The press is regulated by the 1964 Press Law which calls for a 15 man 'press institution' to take responsibility for each newspaper. These people must be acceptable to the Ministry of Information. A publication must have SR100,000 as initial capital investment and a professional staff. Opposition to the ruling regime or its policies cannot be published, not can criticism of friendly countries. On the whole the press is a bland account of events, both internal and external, and the foreign language press is designed to fill in resident foreigners on world political and social events. The imported foreign press is subject to severe censorship to ensure the removal of all offending material, such as the mention of the word Israel. The result is papers riddled with smaller or larger holes.

The conflict and contradiction between westernisation and the maintenance of traditional Wahhabi values is most acute in the development of radio and television. The often recounted story of how Abdel Aziz forced the 'Ulema to accept radio as a religiously permissible phenomenon although it is not mentioned in the Koran, shows the way in which the Saud can manipulate religion for their own ends. In the face of the 'Ulema's oposition to the installation of a radio transmission station, Abdel Aziz had parts of the Koran transmitted to the waiting 'Ulema, and then pointed out that anything which transmits the word of Allah could not be against the faith. Having thus won the argument, radio transmitters were installed — and used largely for communication between the various royal palaces in the 1930s! It was only in 1949 that radio for the lesser mortals was introduced and a public broadcasting system started. Even then a low powered medium wave transmitter was used and broadcasts lasted only a few hours daily, composed entirely of news, religion and other talk, but excluding music, female voices and any form of entertainment, all of which were forbidden by religion. This minimal service operated only in the western region, under the strict control of the regime.

It was political pressure which persuaded the regime to expand and liberalise the media services. Cairo's Voice of the Arabs could be heard throughout the Arab world and was widely listened to in Saudi Arabia. These programmes had considerable popular appeal thanks to their style which contrasted sharply with the tedium and puritanism of Saudi Arabia's radio. Outide the Hejaz, the Voice of the Arabs was the only station which could be heard. To counter this subversive influence, transmitters were built and broadcasting stations set up to cover the entire kingdom by the mid-sixties, allowing for longer hours of programming. Music and women's voices were introduced to retain the interests of listeners and, of course, the opportunity was taken to relay the regime's propaganda. There are currently two radio stations: one in Jeddah and the other in Riyadh, broadcasting 20 hours daily internally and also transmitting an overseas service.

The pressures which forced the regime to improve and liberalise its radio services also applied to the creation of television. Although television was first introduced in the eastern region by Aramco in 1957, the regime showed no signs of developing its own Arabic service. It was only in the early 1960s when efforts to divert public interest from the Voice of the Arabs was a prime objective, that the Saudi regime decided that the best way to keep people away from their radios was to give them a new and better toy: television.

It is interesting to note that a country which still forbids public cinemas (although most rich people now have private projectors and import films and video from abroad) should have allowed the introduction of television. The expressed opposition to cinema and television from religious quarters is the same: that human images must not be represented as this violates Islamic principles. This argument obviously holds as true for TV as for films, but opposition to TV was quashed for a number of reasons: first it is controllable by the government as it is centralised; secondly the neighbouring states had it although they were poorer and it was therefore a matter of status for the regime not to be left behind; lastly the people wanted some entertainment and were spending far too much time listening to Cairo.

In 1963 the regime asked for US cooperation and advice in the selection of a television system, thus ensuring that an American system would be chosen, and starting a profitable rip-off for the US Corps of Engineers, RCA, NBC and AVCO who controlled Saudi Arabian television until the early 1970s. In that time contracts changed hands and dissatisfaction grew. Discussion over the installation of colour television in the 1970s led to an agreement in August 1976 with France for the installation of a colour network by 1981-2. The shift from US to French television equipment is hardly surprising since it was discovered that the Americans had sold the Saudi Arabians obsolete experimental voice transmission equipment 'adapted' for television before shipment to Saudi Arabia. The station which had been commissioned for Dammam and was meant to provide year round transmission to Kuwait only worked on the few days a year when the weather was suitable!

Transmission was started in late 1965 reviving once again conservative religious opposition. In September 1965 a group of opponents of television attacked and destroyed the Jeddah transmitter station and in the battle their leader, Prince Khaled ibn Musa'id was killed (his brother Faisal subsequently killed King Faisal in 1975). Despite this, television broadcasts started successfully but extremely strict rules are observed to minimise religious opposition.

Women did not at first appear on television and today only foreign women are seen as singers or entertainers. The general content of Saudi television can be seen by the following list of unacceptable items: scenes which arouse sexual excitement; women who appear indecently dressed, in dance scenes, or scenes which show overt 'acts of love'; all 'immoral' scenes; women in athletics, games or sports; alcoholic drinks or anything connected with them; derogatory references to any of the 'heavenly religions'; treatment of other countries with praise, satire or contempt; reference to Zionism; material meant to expose monarchy; reference to betting or gambling; excessive violence.

The effect of these criteria on an average imported programme can be seen in the following treatment of a western in the early days of television:

> 'The town sheriff walks into a bar — censored because alcohol is forbidden. Sheriff talks to woman who is unveiled — censored because woman's face is shown. Sheriff pets dog as he walks down the street — censored because the dog is considered an unclean animal. Finally all scenes involving the sheriff are omitted because it is discovered that the sheriff's badge closely resembles the Star of David, and is

unacceptable because of the association with Israel.'[8] Although television is an extremely popular medium today, and watched by about two-thirds of the population, its programming is still limited to Egyptian soap operas, children's programmes, religion and carefully edited news and current affairs.

The period of King Faisal's reign was marked by consolidation and stabilisation of the regime. A national administration was set up and the forms of political control were derived from the tribal system on the one hand and from imported repressive structures on the other. Although Faisal was primarily responsible for the stabilisation which took place, his policies were influenced by the internal and external conflicts he had to deal with in the early days of his reign. It is only under King Khaled's rule that the regime has started to relax the repression of opposition, although it has not introduced any form of democratic consultation of the people.

The introduction of state structures has not yet resulted in a significant weakening of the regime's reliance on religion as a means for ideological control. On the contrary, religion remains one of the main elements in the battle against political change. The regime hopes that its influence among youth will make it possible for Saudi Arabia to develop and retain a strong national identity, which has only recently been achieved. At the moment this identity appears to some to be in danger of dilution as a result of the westernisation which constantly affects daily life. At another level westernisation is associated with political liberalisation which has few powerful supporters, but is all the same a focus of interest to a population which has no power within the state.

8. Douglas Arrington Boyd, *An historical and descriptive analysis of the evolution and development of Saudi Arabian television 1963-1972*. Unpublished PhD, 1972, University of Minnesota, p.242-3 for the quotation and list of criteria.

CHAPTER V
POLITICAL OPPOSITION

An understanding of the nature, type and strength of the opposition in Saudi Arabia requires an awareness of the political context in which it operates. Political opposition to the royal family and to monarchical rule is forbidden. This means that no political parties are allowed and no disagreement with the regime's policies may be published in the press, broadcast, or disseminated in any form. Trade union organisation is illegal and severe prison penalties are imposed on any worker who organises or participates in a strike or any other form of collective industrial action. When joining the United Nations in 1945, the regime refused to sign the Universal Declaration of Human Rights on the grounds that the Koran covers all that is necessary. Civil liberties are non-existent and overt political activity is rewarded with imprisonment without trial.

The Saudi regime has not only had ready at hand the traditional means to suppress opposition, it has also created new instruments as the need arose. An example of this is the promulgation of the State Security Law in March 1961, which prescribes the death penalty or 25 years imprisonment for any person convicted of an aggressive act against the Royal Family or the state. This includes treason and any attempt to change the regime or spread disaffection among the armed forces. In the 1960s, imprisonment, torture and, in the case of aliens, deportation, were the order of the day. Naturally no figures are available.

The intention and the effect of these policies is to spread fear and keep people in ignorance of alternative possibilities of political life. Given these conditions any person with the courage to stand up for his/her beliefs, and to oppose the regime either individually or collectively deserves respect.

After the first ever strike, in 1931, of Abdel Aziz's drivers was repressed and the strikers beaten and deported, there was no further manifestation of unrest for many years. In the 1950s strikes took place in the oil industry and in the 1960s some minor labour disputes occurred in different sectors. Political opposition took the form of demonstrations on the occasion of major events in pan-Arab affairs, and a number of attempts at urban terrorism on the other. At another level, political discontent among the elite was visible in the attempts to develop machinery for consultation and in the conflicts over policy among the royal family.

Opposition within Ruling Circles

Given the size of the ruling family (estimates vary between 2,000 and 7,000 adult princes) it is only natural that different political tendencies should emerge, and that their supporters should struggle for power. Despite the regime's attempts to supplant tribal loyalty by loyalty to the 'nation' as represented by the Saud regime, there are still tensions between tribes and particularly between members of the ruling family whose maternal affiliations associate them with the less powerful factions.

Abdel Aziz's policies of marrying into the tribes he had conquered mean that today, for example, there are Princes who descend from the al-Rashid family of the Shammar who were ousted from Najd by Abdel Aziz. King Faisal's assassin was of al-Rashid descent.

The Free Princes

The only occasion on which there was 'progressive' opposition from members of the royal family was in 1960 and the following years, when Prince Talal ibn Abdel Aziz, under the influence of Arab nationalism, made some proposals for Constitutional reform. King Saud who at that time was trying to reassert his position of dominance over his brother Faisal, the Prime Minister, found it opportune to use Talal's support as well as that of other Princes who objected to Faisal's austerity measures and consequent reduction of their vast allowances.

In December 1960, Talal was appointed Finance Minister and a Constitution announced. This initial success was shortlived: within three days, Mecca Radio denied that the Council of Ministers had approved any such resolution, and it soon became clear that the King had no intention of reforming the political system. Prince Talal was dropped from the Cabinet in September 1961 when he stated that the Dhahran base was to be liquidated within eight months and the agreement with the US over it abrogated, and repeated his call for a constitutional monarchy. By the summer of 1962 Talal's relations with the regime had deteriorated further and in August he gave a press conference in Beirut in which he protested at the occupation of his palaces in Riyadh:

> 'The fact is that I know of no reason why this vindictive measure should have been taken, apart from those suggested by certain papers and radio stations: namely my visit to Egypt and my cable to President Nasser (congratulating him on the successful launching of Egyptian-made rockets)... As regards reports published by some newspapers concerning the freeing of my slaves and concubines, they are quite unfounded because I don't have any slaves or concubines to free. But I will say that I am opposed to slavery...' He continued to state that 'Our aim is to establish a constitutional democracy within a monarchical framework.'[1]

As a result of this press conference his relations with the regime were broken, his passport was withdrawn, his property confiscated, and some of his supporters in Saudi Arabia were arrested. He then moved to Egypt with his supporters from the royal family, his brothers Abdel Mohsen ibn Abdel Aziz, Badr ibn Abdel Aziz, Nawaf ibn Abdel Aziz and his cousin Sa'ad ibn Fahd. While in Egypt, they adopted the name of the Free Princes and broadcast progressive statements on the Voice of the Arabs, where they spoke on behalf of the *Saudi Liberation Front*. The Free Princes issued anti-monarchist propaganda but objected to calls for the massacre of the Saudi royal family broadcast on Sana'a radio. Their relations with the Egyptians soon deteriorated particularly as the Nasserist radio authorities insisted on censoring the programme the Free Princes broadcast on 'Enemies of God'.

By the summer of 1963, Prince Talal was asking to have his Saudi

---

1. Gerald De Gaury, *Faisal*, London 1966, p.104-5

passport returned, and in early 1964 the Princes returned to Riyadh, having made suitable amends. However, by the time Talal left Cairo he owned the highest tenement building and the biggest cinema in the city as well as three palaces. Today the Free Princes are well integrated into the family structure and in 1977 Talal's name reappeared in the press, but for a different reason. He has become a playboy owning a plane decorated with gold and losing money at the gambling tables of the world, and in 1977 was negotiating for the production of a film about Abdel Aziz, for which he has written a script.

Consultation and conflict within the ruling elite

Although the first suggestion for a constitutional monarchy was not taken seriously, the idea keeps being mentioned, thus revealing that there clearly is some demand for it among the educated technocrats and other influential people, such as the trading bourgeoisie. When Faisal finally came to power he promised a constitution (see above page 63) and after he died, the creation of a consultative council was once again mentioned. But nothing has ever been done to implement these proposals, and the mention of any form of consultation usually arouses considerable hostility from members of the ruling family. The 1926 Hejaz constitution was supposed to extend to the rest of the Kingdom, but never did. The 1962 Ten Point Programme's constitution was ignored and the promises of 1975 are similarly forgotten.

But despite this, it must be said that people who are not members of the royal family are now exerting some influence on state decisions. This as yet only applies to the elite of higher technocrats who have senior positions in ministries and to the trading bourgeoisie. It is also true that these groups of people on their own have little or no power, and it is only when they are in alliance with a powerful segment of the ruling family that they are able to exercise any influence. For example in the debate between the modernists and the traditionalists in the late 1970s, there seems to be an alliance between the traditionalists led by Khaled and Abdullah, and some of the technocrats such as the Minister of Planning Hisham Nazer, who favour a slower rate of development, and greater control by Saudi Arabians of the resources of the country and its future policy options. It is unfortunately still the case that without such an alliance with some group in the royal family, no one has any influence on policy decisions, even if there is a large body of people who hold the same opinion. The crucial factor in power is the connection with the royal family.

As the case of the Free Princes demonstrates, opposition to the regime within the ruling circles is fickle. The Princes themselves have now been reintegrated into the family and bought off. They no longer appear to have any interest in politics. Their failure is to some extent due to the fact that they took their opposition outside the usual channels of 'court' struggles for power. This has not been done since. The royal family, despite internal differences on policy, have in the past retained a front of unity, as its members are well aware that to reveal their divisions would open the road to their overthrow. Despite this, differences are sometimes visible and in the late 1970s the main one is between the traditionalists and the 'modernizers'. Any future split within the family is likely to be very different and far more serious than the problems brought about by the Free Princes and awareness of this is one factor leading to compromise between the existing factions. For unlike most states where the force of arms is monopolised by the state, in Saudi Arabia, each of the two main factions has its own military arm, the army on the one hand and the National Guard on the other. Serious and open conflict between these two factions could therefore lead to a violent breakup of the 'state' which would harm both groups. The likelihood, therefore, is that the balance between them will be more or less maintained, and no attempt made to oust one group totally. The closing of ranks in the royal family which took place immediately after King Faisal's assassination indicates the family's awareness of the dangers of division and although conflict between the two branches was brought much closer by Faisal's death, it is likely that the Council of Senior Princes will ensure the continuation of some balance in the succession, which means preventing the Sudairi brothers from monopolising power by strengthening other groups in the family. A start towards this extension of power within the family was effected by the introduction of younger members of the royal family to cabinet posts in the October 1975 government reshuffle.

Left Wing Opposition

During most of the 1950s and early 1960s the regime in Saudi Arabia was in a state of conflict with opposing forces. The 1950s were marked by a rapid influx of western technology, goods and personnel which, on the one hand, challenged the traditional values and, on the other, increased the regime's income without improving the conditions of survival of the majority of the population. Thanks to the level of conspicuous consumption engaged in by the royal family, the people could not fail to notice the luxury in which the royal family lived, and also became acutely aware of the fact that they were getting no share of benefits from this manna. Everybody wanted transistor radios and modern utensils, but all the people got was the sight of luxury palaces, vehicles and luxury goods being wasted on an unprecedented scale.

These very visible changes emphasised the degree of US influence on the regime at a time when the majority of the people in the Middle East were becoming particularly conscious of the meaning of imperialism as a result of the 1948 partition of Palestine, and the support given to Israel by the West. In Saudi Arabia, Palestinian refugees started arriving and obtaining employment in the oil industry, and the indigenous population was reminded of President Roosevelt's statement to King Abdel Aziz, assuring him that no decision would be reached on Palestine without consultation of the Arabs. This pledge had been broken without so much as a 'by your leave' by President Truman, whose interest in Saudi Arabia was restricted to ensuring the flow of oil and money to American coffers.

Popular awareness of Arab nationalism in the broadest sense started to grow in the early 1950s in Saudi Arabia under the influence of exiled Palestinians and the increased awareness of the connection between the US presence and the development of problems in the internal situation.

The emergence of the *Movement of Arab Nationalists* which developed in the early 1950s in Beirut under the leadership of the Palestinians had a theoretical influence on nationalists throughout the Arab world. In 1952 came the Egyptian revolution and Nasser's subsequent conversion to nationalism, as well as to pan-Arabism. He gained enormous prestige throughout the Arab world in 1956 for his stand for the nationalisation of the Suez Canal and against US and British imperialism, and a massive upsurge of nationalist enthusiasm swept through the Middle East. All sections of the population were affected as, for the first time, it became clear to the vast majority of Arabs that subservience to the west was not an unavoidable fate. Saudi Arabia was not immune to the excitement, and the newly educated young administrators and intellectuals were revolted by what they saw as a totally corrupt government under the influence of any Western capitalist who cared to offer a them a new gadget at inflated prices. A widespread progressive movement developed rapidly in the country, manifested by strikes, discontent in the army and the emergence of a number of nationalist organisations.

The main grievances of the young progressives were the absence of 'modernisation' in Saudi Arabia, the country's subservience to the west and in particular to US imperialism, the lack of democratic rights, and the corruption and waste within the regime. They felt concerned at the apparent 'backwardness' of the country which they compared to the USA, where they had studied. Differences were soon clear between the progressives whose outside experience was Arab and those whose experience was western. The latter were sufficiently satisfied by the changes made by Faisal to continue working within the regime, whereas the former were not because of the poltical atmosphere. The opposition groups which emerged in Saudi Arabia in the mid-1950s were at first organised on only a local basis. Many of them consisted of immigrant Yemenis, Egyptians and Palestinians who had flocked to Saudi Arabia as the oil boom demanded professional, skilled and unskilled manpower. They had considerable political influence on the Saudi Arabian workers in the oil fields.

The 1953 and 1956 Aramco strikes

As the only sizeable employer in the country, Aramco's treatment of its workers was seen by Saudi Arabians as a reflection of American attitudes to them. In the 1940s and 1950s the company's discriminatory social and economic policies meant that although, in absolute terms, the standard of living of the company's Saudi Arabian employees was rising dramatically, their experience was one of humiliation. The company had separate living quarters for Americans, intermediate staff and Saudi Arabians, whose conditions ranged from luxury villas to army-style dormitory barracks; in office blocks, lavatories were segregated according to the rank and nationality of the employees. Such practice did not endear the company to the local population, who felt like third class citizens in their own country. In 1952, although Saudi Arabians formed 61.7 per cent of the labour force, 69 per cent of them were unskilled and only 6 men were in professional or higher level supervision.

The intermediate staff were largely migrant workers, particularly

Palestinians who had been expelled from Palestine in 1948 and found employment in the oil industry, as well as Indians and Pakistanis who also brought a politically experienced perspective to labour relations, new to the Saudi Arabians.

In the early fifties, some of the first men who had been sent to the USA for training by Aramco were returning to a higher employment grade. Despite this they still found themselves living in the worst conditions in the Aramco quarters and therefore recognised the Aramco policy as segregation based on race rather than on employment status.

The first trouble broke out in the summer of 1953 when a number of newly trained Saudi Arabians, acutely aware of the discrimination to which they were subjected, demanded pay increases, better conditions and an end to the segregation of living quarters. The workers presented a petition including their demands to the company, which passed it on to the King. The King's response was to set up a Royal Commission of Inquiry, chaired by Crown Prince Saud which, after a confrontation with the petitioners, ordered that their leaders be jailed. This resulted in a strike which lasted for two weeks and was supported by all Saudi Arabian workers. The Strike Committee members were later exiled to their home villages, but within a few years were back at work in Aramco, as a result of the company's desperate need for workers.

As a result of the strike King Saud (who had since inherited the throne) visited Dhahran and decreed some improvements in working conditions:

> 'All Saudi workers were to receive automatic pay increases ranging from 12 to 20 per cent of their present wage rates. Vacations for Saudis were doubled, Housing for Saudis was to be improved. Job requirements were to be made easier for Saudis. The company was to pay 20 per cent of the cost of family houses built for Saudis under its home loan programme; company schools were to be expanded and the company would also build and maintain ten schools in nearby towns for the sons of Saudi employees. Other benefits ranged from subsidised food and half-price clothing to 550 new water coolers for Saudi quarters. And new systems were set up for handling the grievances of Saudi employees.'[2]

2. M.S. Cheney, *Big Oilman from Arabia*, London 1958, p.271

These concessions showed that Aramco had to give in to some of the demands, but they also strengthened relations between the government and the company at the expense of the workers. Both Aramco and the Saud ruling family hated workers' organisations equally and found it in their interest to co-ordinate their policies of repression of labour demands by adopting an attitude of paternalistic protection: improving conditions on the one hand, and enforcing strong repression against dissidents on the other.

The first strike had been brought about primarily by discontent with local working and living conditions, and had no political aims. The government's reaction, particularly jailing the strike committee leaders, meant that the workers felt the need in future to have a *Workers Committee* which soon developed into a forum for nationalist discussion.

In the years following 1953, discontent simmered just below the surface, to the increased concern of the authorities who retaliated by arresting Saudi Arabians whom they suspected. Such people were imprisoned without trial or right to appeal and spent varying lengths of time in the unsalubrious jails of the regime, and, as far as the world was concerned, had disappeared, as no news of them could be obtained.

In 1955 about one hundred Palestinians employed by Aramco were suddenly 'bagged in a midnight round-up by the local cops and deported, without trial of course, as being members of a pan-Arab political movement known as the PPS, which was unpopular with national governments. Another batch was later expelled on the grounds that they were Communists'. As the author, himself an Aramco employee who makes no claim to being progressive, points out himself, 'there was little, if any, evidence to support this contention in most cases; but being men without a country, they made convenient scapegoats for nervous officials.'[3]

The 1954 working conditions still fell far short of the needs and aspirations of the Saudi Arabian workers who, since then, had been discussing politics and had developed a more consistent opposition to the regime. In 1956 their nationalist feelings were exacerbated by the government's avowed intention of renewing the lease which gave the USA the use of Dhahran Air Base for its forces. This and the dissatisfaction at work led to a demonstration:

3. *ibid.* p.284

'In May 1956, while His Majesty was eating a stuffed lamb dinner Aramco had prepared for him in Dhahran, thousands of workers held a demonstration and in front of the King himself raised banners saying "Down with American Imperialism", "We want an elected Trade Union", etc. Immediately the regime's henchmen including the Eastern Province Governor, Saud Ben Jiluwi, and Director of Investigation Ali Ghamidi, took charge of the assault on the working class. They arrested large numbers of workers and nationalists, who were later subjected to savage physical and mental torture. Still later they were transferred to the underground prison *Sejin alabeed*, the "slaves' prison". Again others, who were being pursued by the regime, fled the country.

'The situation worsened and class antagonism intensified. On June 17 the workers announced still another strike which lasted for several weeks. This strike interrupted the company's work and forced Aramco to give in to some of the workers' demands such as small wage increases and some improvements in living conditions.'[4]

King Saud, who at that time was trying to appear nationalist and progressive, was outraged by the strike and quickly decreed a law forbidding participation in any strike under penalty of imprisonment, and promised even greater punishment for incitement to strike.

After 1956 Aramco adopted a progressive labour policy, giving workers better conditions in health insurance, interest-free loans to build their own homes and other facilities. After the strikes, the company also extended its training programme to create a more stable labour force. This was partly meant to defuse any future labour unrest, as the shift from economic demands in 1953 to political demands in 1956 had greatly disturbed both the company and the government. The company hoped that its new policy of giving particularly good conditions to Aramco workers would pay off in a consequent reduction in political activity.

Political opposition

The *Workers' Committee* set up after the 1953 Aramco strike was

---

4. from *Jazira al Jadida*, translated in *Struggle, oppression and counter revolution in Saudia Arabia*, Berkeley USA, p.5

POLITICAL OPPOSITION 99

the first 'organisation' in which Saudi Arabians met and developed a political perspective. Out of this group arose a number of political organisations, whose significance is difficult to estimate, given the difficulties under which they were operating and their inability to distribute information. But it is clear that they were most active during Saud's reign and in the early years of Faisal's.

Their activity cannot be explained simply by the boost to nationalism provided by Nasser, but also by the state of flux within Saudi Arabia at the time. King Saud's vacillation between support for imperialism and Arab nationalism, the open struggle for power between Saud and Faisal after 1958, and the instability created as each sought allies both within and beyond the royal family, gave greater opportunity for open political activity as there was a chance that one or other would support certain reforms and object to the imprisonment of its advocates. This situation of flux prevailed into the early years of Faisal's reign. Added to this, until 1962 the security forces were primitive and limited. It was only after the beginning of the Yemen War and with the intensification of the internal conflicts this produced, that the regime strengthened the National Guard and brought in US intelligence experts to cope with the growing tide of left-wing activity.

Despite the 1961 State Security Law, the existing progressive groups such as the *Sons of the Arabian Peninsula*, and the *National Liberation Front* continued to send and distribute leaflets in Saudi Arabia denouncing the royal family's extravagance and its pro-western policies, and asserting the people's right to ownership of the land and its resources, including oil.

On a tour of the Eastern region in 1961 King Saud found open opposition from hostile crowds, demanding work and educational facilities; during this tour a triumphal arch was mysteriously burnt down!

The most active period for the opposition started in late 1962 when, after the 26 September Revolution in North Yemen, the Saudi regime decided to support the Imam in his struggle to get back to power and Nasser's troops moved in to support the republican regime. Nasser had made no secret of his ambition to overthrow the monarchy in Saudi Arabia, and he was prepared to subsidise any nationalist groups in Saudi Arabia which vaguely agreed with him. Such groups were supplied with weapons, printing and distribution facilities and access to the powerful Voice of the

Arabs radio. Although at first this was monopolised by the Free Princes, soon other organisations used it. The influence of this radio on the development of anti-imperialist political consciousness throughout the Arab world must not be neglected.

Within two weeks of the Yemeni revolution four Saudi Air Force planes and their crews had defected to Egypt. Sent to Najran, near the Yemeni border, to support the Imam's forces, they flew to Egypt instead, where they announced the existence of an organisation of free officers and civilians in Saudi Arabia.

In late 1962 and early 1963, Egyptian aircraft dropped packs of weapons into the deserts of Saudi Arabia. The weapons included American, British and other arms no longer used by the Egyptian army. At the same time progressive Saudi Arabians were appealing to their compatriots to rise against the monarchy, and eliminate members of the royal family and other agents of imperialism.

> 'The people must revolt after having heard the voice of truth rise in Sana'a, Baghdad and Damascus. This revolt is a sacred task which the people must accomplish by eliminating this cowardly minority made up by the governing class and its lackeys. You are many, you are courageous, and with the strength given you by God and his trust you must make good triumph over evil.'[5]

Despite reassurance from American officials, the regime was concerned lest the US abandon them in favour of the Republican regime in the Yemen, particularly when the US recognised the YAR in December 1962. Saudi Arabia then started to purchase modern weapons and military training from Western Europe and particularly from the UK with whom diplomatic relations had been re-established in early 1963.

Agitation continued in the next few years with occasional reports of defections to the YAR or Egypt, repressive measures taken against foreigners, usually Palestinians, Yemenis or Egyptians, and rumours of attempted coups. 1966 was to prove a year of intense political unrest. Early in the year 19 'communists' were sentenced to between 3 and 15 years imprisonment, while another 34 were given royal pardons after having admitted their guilt and asked for forgiveness. But no details were given of the actions of which they were accused. Whoever these people may have been, their

---

5. *La Bourse Egyptienne*, 10.3.1963

imprisonment in no way halted opposition. In May leaflets were distributed in Mecca and Riyadh announcing the launching of underground resistance by the *Society for the Liberation of the Holy Soil*. Later that month arrests were made and some Egyptians and Palestinians expelled.

In December the *Union of the People of the Arabian Peninsula* distributed many leaflets throughout the country and explosions occurred in Riyadh, Dammam and both Jizan and Najran near the Yemeni border. Evidently well-planned, the explosions took place in the Riyadh main post office, the Ministry of Defence, the office of the senior US advisor in Riyadh, the Palace of Prince Abdel Rahman, the security forces headquarters in Dammam and other military establishments, as well as along Tapline. As a result a number of Egyptians, Palestinians and other Arabs were expelled, the Director of Police in Riyadh was retired, and 15 army officers sacked.

In early 1967 between 400 and 750 people were arrested and in mid-March 17 Yemenis were publicly executed after having supposedly 'confessed' responsibility for the explosions. Their execution was widely condemned in the Arab world and the UPAP proclaimed the innocence of the victims on the grounds that it had been responsible for the explosions. It promised more of them in future. In the days following the executions the Saudi authorities rounded up about 35,000 Yemenis and deported them.[6] The Saudi Arabians who were arrested were never brought to trial but were 'interrogated' and tortured by Saudi, American and British personnel in Riyadh and in the Eastern region. Those who survived the prison conditions were finally released in 1975 in the amnesty declared by King Khaled on his accession.

While the population was still living under the shock of these events, the June War of 1967 was fought and widespread demonstrations took place. These protested simultaneously at US support for Israel, the Saudi regime's refusal to cut off oil to imperialist countries, and the recent imprisonments and executions in Saudi Arabia itself.

> [In Dhahran] 'the demonstrators attacked the American Consulate, took down the American flag, tore it up and burned it. Afterwards they marched to the airport and military

---

6. *New York Times*, 29.3.1963

installations where they destroyed the American's clubs, cars and recreation centres. They also attacked the American military barracks which were symbols of the country's foreign domination. Finally they were stopped by police who prevented them from entering the military airport. Before being forced to leave, the demonstrators set fire to more cars and both occupied and damaged the local Aramco headquarters.

'These mass demonstrations were followed by a new wave of terror and arrest campaigns encompassing the entire country which victimised workers, soldiers and other progressive elements. The traitor Fahad ben Abdulaziz who was Minister of the Interior and the Second Vice-Premier of the Council of Ministers, went to the American consulate and to Aramco's president Thomas Barket to apologise for the workers' action. He also began supervision of the interrogation and torture of people who had been arrested as suspects for participation in the demonstrations. He was aided by the King's henchmen such as the Eastern Province's Governor, Abdul Mohsin ben Jiluwi, and the Director of Investigation.'[7]

In 1969 there were two attempted coups. The first in June was organised by Air Force officers who were supported by some left-wing teachers and Aramco workers. It was defeated by an infiltrator who warned the government only hours before the coup was due to take place. Some officers who were studying abroad were brought back under various pretexts and were arrested on arrival. They were dealt with brutally: some were reportedly thrown out of flying aircraft and many others imprisoned without trial. The coup was attributed to the *Movement of Arab Nationalists*. At the time it was estimated that about 200 people had been arrested.

The second coup, which never came as near success as the first, was organised by some bourgeois reformers from the Hejaz whose families had always opposed the Saud. Originally aligned with reformist Prince Talal, they were promised reforms by Faisal when he came to power and only decided that a coup was necessary when no progress had been made for some years. Their leaders included retired generals Ali Zein Abdine, Abdullah Aysi, and other senior

---

7. *Struggle, oppression and counter revolution in Saudia Arabia*, p.6-7

military officers, under the influence of Ahmad Tawil a close friend of Prince Fahd's. Some members were arrested in July and taken to Dhahran for questioning by US intelligence experts, where they confessed and denounced all their accomplices.

Fear of possible coups continued and in 1970 further arrests took place, including senior Petromin officials and the Dean of the Petroleum College, Salah Ambah, Aramco employees and a senior official at the Ministry of Information. After the September massacre of Palestinians in Jordan in 1970, the Qatif area was sealed off for six weeks under the pretext of a cholera epidemic. While no-one was reported to be suffering from the disease, many arrests took place among the local population, who are Shi'a, and their year's crops were lost as they were prevented from harvesting them.

Since then there have been few reports of incidents, but it has always been very difficult to obtain any information on opposition activities as the regime has actively suppressed any such publicity, going to the lengths of buying up copies of papers which publish hostile reports, particularly in the days of the Lebanese free press.

In May 1977 a serious fire occured in a pipeline near Abqaiq which was described as an accident by the oil company. But there were suggestions of sabotage, and a previously unknown organisation, *Arab Destiny* claimed responsibility for the fire in a communiqué addressed to the Agence France Presse.[8] Although there may have been good reasons for sabotage, if any occured, it is almost as likely to have come from people wishing to demonstrate to the Saudis that they still needed the company, at a time when negotiations for Saudi Arabian control of Aramco were dragging on.

Left wing organisations

As mentioned above, the first ever progressive 'organisation' was the *Workers' Committee* which was formed during the 1953 Aramco strike and continued underground until the 1956 strike; it was severely repressed in 1956 and many of its cadres were arrested. It was a broad front politically, including members with varying degrees of political consciousness and differing political opinions. After 1956 its remaining members disbanded and joined other organisations in which they tended to form the nucleus as their

---

8. *Le Monde*, 18.5.1977

political experience was greater.

In 1956 the *National Reform Front* was formed by a group of communists and former members of the *Workers' Committee*. In 1958 the *National Liberation Front* was created by the communists from the *Reform Front*. This organisation remained weak, but participated in the attempted coups in 1969. Its stated objectives are:

> 'Far reaching change in all aspects of Saudi Arabian life. We are for a state system that would speak for the people's interests and pursue a policy against imperialism, zionism, and reaction. We demand a democratic constitution ensuring basic rights, including the right to set up political parties, trade unions and other mass public organisations, the right to strike, hold demonstrations, meetings, etc... We demand the dismantling of all foreign bases in our country and the abrogation of the shackling military agreements forced on it. The Front also wants a revision of existing concession contracts with foreign oil monopolies to provide for the principle of broad state participation in the entire process from prospecting to marketing... The nation is in urgent need of a public sector of the economy... The Front stands for extensive political relations and close economic and cultural co-operation with the Soviet Union and all countries of the socialist community.'[9]

In late 1975 the *National Liberation Front* announced its change of name to the *Saudi Arabian Communist Party*.

The most active organisation in the 1960s was the *Union of People of the Arabian Peninsula* (UPAP), a Nasserist organisation led by a former Aramco worker, Nasser Said. The Union claimed responsibility for the bombs which exploded throughout the country in late 1966, and issued statements on the Voice of the Arabs and distributed literature in and out of the country throughout the decade. In recent years they have published no programme and seem to have been much weakened by the 1969 arrests as well as the ideological decline of Nasserism. Their support is to be found mainly in the army and among the Shammar tribe in the North, who have traditionally been opposed to the House of Saud. Their appeal is mainly emotional, based on demands for Arab Unity and a 'better' society, with greater

---

9. *Peace, Freedom and Socialism*, November 1974

benefits for the poor.

The Saudi Arabian branch of the *Ba'ath Party* was founded in 1958. By 1963 it was the largest opposition group but split in that year between the Syrian and Iraqi lines when the Ba'ath party quarrelled at the Arab level. Many of its members then left the party altogether, those remaining being mainly supporters of the Syrian Salah Jadid line. However, as their membership was mainly in the army, most of them were arrested in 1969. Currently there are small groups of both factions: the pro-Iraqi faction is clearly the most active and publishes a regular magazine *Saut al Tali'a*.

The *Popular Democratic Party* was founded in 1970 after a series of meetings between former members of the *Movement of Arab Nationalists* who had been active in Saudi Arabia since 1964 and had spread their activities to the west and centre of the country by 1969. In 1970 they decided to form a party composed of independent Marxists and nationalists with Marxist leanings. They consider themselves Marxist-Leninist and hope to liberate the country by armed struggle. Their support comes mainly from students and petty bureaucrats. They are the first organisation to have a women's branch, and four of their women were among those arrested in 1969. Like all other groups they lost many members in the 1969-70 arrests. Their aims are to bring down the regime, obtain full civil rights for all people, and particularly for women, and to defeat imperialism old and new. They want an economic policy of planned development of agriculture and industry, the formation of a revolutionary government, and support all socialist and revolutionary movements. At present they believe in the need to unite all people against the regime by means of the establishment of a national front. The party has an irregular magazine called *Al Jazira al Jadida*.

In 1971 The *Popular Struggle Front* split from the *Popular Democratic Party* and published a paper in the US called *al Nidal*; they are believed to be an insignificant force.

All these organisations are currently extremely weak and have very few members in the country. They are illegal and therefore cannot publicise their activities or ideas, and the penalty for membership of any of these organisations is death. Their influence among Saudi Arabian students abroad is somewhat greater, but even that is often temporary, as the students return to large salaries and a secure position in Saudi Arabian society.

There has been no evidence of organised or even spontaneous political activity among nomads or peasants who continue to use the traditional tribal channels when they have demands or grievances.

Why has the opposition failed?

Monarchies have been overthrown and progressive regimes set up in various Middle Eastern states. This has not happened in Saudi Arabia. The reasons can be found both in the objective situation of Saudi Arabia and in the weaknesses of the left.

Having been united so recently, the country still suffers from a lack of national cohesion at the political and social levels. This lack of cohesion is due to the persistence of old problems: the low population density (O.35 people per km^2) must be added to the dispersal of the population in a number of regions which have historically had considerable autonomy due to their geographic isolation from one another. People therefore continue to identify on a regional basis as Hejazis, Najdis, and Gulf people; they feel a greater affinity to their immediate neighbours than to other Saudi Arabians who are on the opposite side of the peninsula and have a different culture and economy. Similarly the economy is still composed of different sectors in the different regions: oil in the east, trade in the west, agriculture in Asir and administration in Najd.

Struggles for change are often based on economic deprivation confronted with extreme wealth. Although in a particularly striking form in Saudi Arabia the regime has the means, thanks to its vast financial resources, to provide the necessities of life and more to most of its citizens, and in this way, it can do much to defuse a considerable degree of dissatisfaction. The traditions of tribal society and individual access to the ruler also contribute to a popular belief in the possibilities of improving personal and family situation by these means.

Arising from the recent unification and the development of a new economy based on the income from oil sales, is the fact that the country is in the midst of an extremely rapid process of transition between a 'tribal' society and an advanced 'capitalist' one. The speed of this transformation has disrupted the traditional social structure, but a new class structure in which each class has its own clearly defined interests which conflict with those of another class, has yet to be formed. This absence of developed class forces precludes the possibility of a Marxist revolution as there is little unity on which to base collective organisation and demands.

Many of the anti-monarchic movements in the Middle East have been military coups, rather than revolutions. Although there have been some attempted coups in Saudi Arabia, none has succeeded. In this context it is important to differentiate between the army in Saudi Arabia and most other armies in terms of their potential for political action. The army as a rule is the armed branch of the state and in the Arab world it has been involved as a political force because, on the whole, it has seen itself as a focus for modernisation thanks to its skills in using sophisticated equipment. It has also considered itself a prime mover for purity in politics, opposed to corruption and in favour of the less privileged strata in society.

In Saudi Arabia, however, there is not one, but two armies. They are kept well apart and rival each other in training and equipment. Each is associated with one of the two main branches of the royal family which are contending for power. Attempted coups have taken place in the Air Force and the Army and the left wing organisations have all tried to develop a strong membership in the armed forces, as in the Middle East it is assumed that the army will be the source for 'revolutionary' action. But no left organisation even claims to have any members in the National Guard which is totally linked with the 'traditionalist' branch of the royal family. Although its stated duty is to defend the regime against any internal opposition, were a battle to occur between the two factions, there is little doubt that it would side with its leaders and fight in support of Wahhabi values and against westernisation. Meanwhile its loyalty to the regime ensures its reliability, and there is no likelihood of it being involved in any coups, unless in reaction to an unlikely liberalisation of the regime.

Just as the Saudi Arabian armed forces are different from others, the Saudi monarchy is unique. Unlike the Egyptian, Jordanian, or other royal families in the Middle East, the Saud are more than just a family. Numbering perhaps 5,000 male princes, the family is a tribe in itself. Thanks to Abdel Aziz's political use of the marriage bed, the family includes members related to all the major tribes in the country, thus creating a wide base of support for the regime. Similarly, thanks to its size, it is a politically powerful force as it can monopolise all the important political positions in the country and still have men to spare. This gives it a potential which no other ruling family has had, and it is making good use of it.

Apart from these structural factors, there are also reasons for the left's failure which derive from its own weaknesses. Beyond the shared feelings of nationalism and dissatisfaction with the monarchy and its relations with US imperialism, the left has been divided. Suffering from the problems of the Arab left in general, it has failed to develop its own analysis of the specific conditions of Saudi Arabia, and instead has taken up political positions at a vague 'Arab' level. Most of the organisations we have mentioned come out of a movement based abroad, and although the Ba'athists, Nasserists, and the *Movement of Arab Nationalists* all claim some kind of pan-Arab ideology, none of them provides a clear theory on which independent national and social analyses could be developed. On the contrary, all base their policies on an emotional appeal to the idea of the Arab Nation.

The left's divisions have often been based not on disagreements over particular points of a strategy for Saudi Arabia, but over issues fundamentally irrelevant to Saudi Arabia, such as criticisms of the Soviet Union's form of communism and the comparative merits of the Iraqi and Syrian Ba'ath.

The left has always responded to events in the Arab world in general and Palestine in particular. They have not *initiated* any actions on the basis of internal needs, objectives and possibilities within the specific Saudi Arabian situation. Apart from anything else, this reflects the absence of concrete analysis of the situation in the country as it affects its native and migrant populations. Like most other left-wing movements in many countries, they have accepted simple models which pass for class analysis and attempted to fit their country into a pre-determined pattern which derives from a European or Asian model with nothing in common with their own society. The *National Liberation Front*'s programme, for example, could apply to practically any country: it shows no awareness of Saudi Arabian specificity, describes the economy as 'semi-feudal capitalist', calls for the creation of a state sector, ignoring that which already exists, and appears to base its international relations on a call for 'closer relations with the Soviet Union'. While the communists abandon analysis for the easy rhetoric of socialism, the Ba'athists have nothing to offer in terms of a development strategy in Saudi Arabia, as they see the Arab world as one 'nation' and therefore do not seek any solutions for any single 'state' within it.

This fundamental inability of the movement to grapple with

internal issues and to raise demands which are of immediate interest to the population is very often true of the Arab left in general, as well as in Saudi Arabia. This can be seen in the left's history in Saudi Arabia; apart from the Aramco struggles in 1953 and 1956 which were largely caused by dissatisfaction with working conditions, the waves of left-wing activity have existed almost entirely as a side-show of other Arab states' foreign policies. For example the peaks of activity in 1962-63 and later in 1966-67 both coincided with a Nasserite onslaught on the Saudi regime, and the movements depended on Egyptian support for both material and propaganda.

After 1967, once Nasser had been forced into subservience to the Saudi monarchy (see Chapter VI below), little was heard of the Nasserist movements and they were too weak internally to initiate any activity. Without external material support the movement appears to have collapsed. There have been no significant reports of disturbances or dissatisfaction in recent years. The existing groups, largely in exile, are only able to publish regularly if they have external backing such as the Ba'athist *Saut al Tali'a* group, whereas the Marxist *Popular Democratic Party* which has no external support, is unable to bring out regular publications. The importance of publications cannot be neglected as they can have considerable impact on the large community of Saudi Arabian students abroad, who will then return home with new ideas.

The difficulties experienced by the Saudi Arabian progressive movement both at home and abroad, call for a period of reflection and deep analysis to produce more appropriate strategies. It seems that some parts of the left-wing movement are aware of this as they are concentrating their energies on study and analysis. The pessimistic conclusion may be that the class and political structure has to complete a process of development before radical action can become effective.

CHAPTER VI
SAUDI ARABIA'S FOREIGN POLICY

In the first years of its existence Saudi Arabia was of limited significance to the world scene. The new state was surrounded by British-protected territory, but having been forced to relinquish its Protectorate treaty over Saudi Arabia, Britain had no interest in making incursions into a barren desert. Saudi Arabia's main link with the outside world was as the centre of Islam: every year it was host to the pilgrims who trekked from all over the Islamic world to the Holy Places, and world Islam took an interest in Saudi affairs only insofar as they affected the upkeep of the Holy Places and the pilgrimage. The regime's interpretation of of its religious duties was in theory a powerful motivating force for expansion, but no attempt was made to extend its control beyond the borders, both because the British stood in the way and because the regime was aware of its own weakness.

In these early years the main focus of interest for the regime was its relations with the Hashemite states of Jordan and Iraq. Deep suspicion and rivalry between Sauds and Hashemites persisted after the Hashemites had been ousted from the Hejaz, and their presence on the throne of two neighbouring states meant that the only land links between the Peninsula and the outside world were under Hashemite control. To Riyadh this was a threat, and although at certain moments in later years the Sauds and the Hashemites found it expedient to ally against common enemies, relations have never been close.

Saudi Arabia enters the world

During the Second World War Saudi Arabia entered the world political scene, and the basis of its future foreign policy was laid. The discovery of oil did not immediately make the country a focus of international interest, but as its full significance became apparent, Britain and the United States struggled for control over the area, a struggle decisively won, as we have seen, by the Americans. As the War had further reduced the numbers of pilgrims (already decimated by the world Depression) the Saudi regime looked for new sources of revenue which were found in the oil concession: the basis for the special relationship between Saudi Arabia and the United States was laid. As the War drew to a close and American superiority over Britain became clear, Roosevelt invited Abdel Aziz to meet him on a United States Navy vessel in the Suez Canal (surely a highly symbolic choice of location!). This was Abdel Aziz's first trip abroad and marked his consecration as a good ally and as a figure of some standing in world politics. After a subsequent meeting with Churchill, he declared war on Germany (in March 1945) thus ensuring Saudi Arabia's right to join the United Nations, though it was not at first an active participant.

After the war, conflict over Palestine came to a head. When the State of Israel was created, Abdel Aziz felt betrayed by the United States since Roosevelt had led him to believe that he would oppose such a move. Saudi Arabia sent a battalion to fight in the Egyptian section of the Arab army, but otherwise could do nothing, since the regime depended on US aid. The hamstrung Saudi policy on the Palestine issue was clear.

The first stirrings of Arab nationalism came out of the war period and the Palestine issue. In 1945 the Arab League was formed, with Saudi Arabia as a founder member, but joining only after a clause had been inserted in the Charter guaranteeing that the League would not impinge on the national sovereignty of member states. This minimal lip-service to the cause was the closest the regime ever got to supporting Arab nationalism.

By the time Saud succeeded to the throne, the main issues in the regime's foreign policies, and its basic approach to world affairs had been defined: alliance with the West, in particular the United States; hostility to Israel, but also a fundamental hostility to Arab nationalism; thus a strategy attempting more or less consistently to balance the over-riding dedication to American imperialism with commitment to its own interpretation of the Arab cause. Saudi foreign policy cannot however be fully understood without bearing in mind that the regime remains dedicated to a highly militant form of Islam.

Foreign policy under Saud

Like his internal policy, King Saud's foreign policy shifted between radicalism and reaction. In his first years he fought Britain in the international courts over Buraimi, flirted with nationalism by advocating neutrality, denouncing the Baghdad Pact, calling for the liberation of Algeria and Palestine, and signing a Mutual Defence Pact with Egypt. He rejected US aid on the grounds that the sum offered was an insult by comparison with the aid to Israel, and made overtures to the USSR over diplomatic relations and arms purchases. These moves either failed or were not brought to any conclusion. There was, for example, never any serious intention of setting up relations with the Soviet Union, nor of booting the USA out of the Dhahran air base. These 'nationalist' policies were partly aimed at proving to the Arab public that Saud was a better nationalist than Nasser. By late 1956 Nasser had gained incomparable prestige by his nationalisation of the Canal and by standing up against the attack of the old imperialist powers, albeit with the partial support of the new one, the USA, and it was clear that Saud could not compete with him on these terms.

The only conflict with imperialism which became acute at that period was the crisis over the Buraimi oasis. Beginning in the last years of King Abdel Aziz's reign, the Buraimi oasis dispute spanned Saud's entire political career and the early years of King Faisal's. Composed of 9 villages, the Buraimi oasis is located in an area where borders had not previously been defined, between Saudi Arabia and British-protected Abu Dhabi and Oman. Of no strategic interest until the discovery of oil, it was then claimed by all sides. The Saudi Arabians, encouraged by Aramco (which believed there to be oil in the area) and thus indirectly by the USA, sent in troops in 1952 to assert their domination over the oasis. These were expelled in October 1955 by the British-officered Trucial Oman Levies, while the International Arbitration Tribunal, which had been accepted by both parties, was meeting in Geneva. This action ended negotiations and a year later, after the Anglo-French-Israeli attack on Suez, the Saudi Arabian regime under King Saud broke off diplomatic relations with Britain. Ostensibly meant as retaliation against Suez, this move is perhaps better understood as a move in the border dispute. In the following years the Saudi regime supported the Omani Imamate with finance, weapons and bases, in its struggle against the British-dominated Muscati al-bu Sa'id dynasty which was trying to gain control over the Omani interior, and eventually succeeded thanks to British military intervention.

Saud's 'nationalist' stand in the mid-1950s alienated the USA and Aramco, whose support was essential to maintain his standard of luxury and his control over the country. As, thanks to Suez, his attempt to lead the Arab world as a nationalist had failed, Saud turned back towards a United States-sponsored leadership. He timed his visit to the USA very well. Arriving in 1957, a few weeks after the announcement of the Eisenhower Doctrine, the King was received gracefully by the Government and the oil companies, though he was severely snubbed in New York, because of Zionist influence. He was persuaded to support the Doctrine, and on his way back to Riyadh visited Cairo where he tried to obtain the support of other Arab leaders for the Doctrine. This failed, and a compromise statement was issued which called on the Arab world to stay out of the Cold War and "abide by the policy of impartiality and positive neutrality thus preserving its real interests."[1] Saud returned to Riyadh and the conflict between himself and the nationalists then developed fully.

In early 1958 Arab states were uniting like mad. On 1 February the United Arab Republic was founded, uniting Syria and Egypt, at that time the only two progressive regimes; and two weeks later Hashemite-controlled Jordan and Iraq formed the rival Arab Federation, which only lasted till the overthrow of King Faisal of Iraq, in the 14 July Revolution five months later. Saud felt trapped: he was unable to join the former because of his pro-American policies, or the latter because of the remaining suspicion between the two dynasties, even though he had received King Faisal in September 1956.

Saud did not linger. In March 1958 he offered the head of Syrian military intelligence, Colonel Abdel Hamid Sarraj, £2 million to assassinate Nasser.

The public outcry caused by Sarraj's revelation of this attempt

1. *The Times* 18.2.1957

ruined what little prestige the regime still possessed in the Arab world. Not only had it tried to dominate the Arabs by pretending to be nationalist while it renewed the US Air Force lease for the use of the Dhahran airbase, and subscribed to a Cold War pro-western policy clearly in contradiction with the aims and objectives of neutrality of Bandung which the regime had accepted, but also the King himself had gone as far as to seek the assassination of the most prestigious and revered Arab leader of them all!

By 1958 Saud was considered an enemy by most Arabs, and a highly unreliable ally by the Americans. The financial crisis, added to the scandal of the assassination attempt, were enough to persuade the Council of Senior Princes to immediately hand over all power to Faisal.

Foreign policy under Faisal

In contrast with Saud's, Faisal's foreign policy was based on a number of principles from which he never swerved. These were: the spread of Islam, and its use to increase Saudi Arabia's influence in the world; a pathological hatred of communism and by extension of anything progressive, since these were associated in his mind with atheism; hatred of Zionism, which in Faisal's mind was closely associated with communism, (this went so far that Faisal even made the absurd statement that 'communism is a zionist creation aimed at achieving the objectives of zionism'[2]); and finally the maintenance of the alliance with the United States, an alliance to which anti-communist fanaticism was no obstacle.

The application of these principles is easy to see in the foreign policy pursued under his control. Having taken over in the midst of a period of conflict with Egypt, during his first years Faisal concentrated on improving Saudi Arabia's domestic situation, whilst managing to maintain cool but cordial relations with Nasser. But by the time the Yemen War broke out and Faisal was once again put in charge of a situation bungled by his brother, he had for years tolerated daily broadcasts about his country entitled 'Enemies of God' on Cairo's Voice of the Arabs, in which the regime was virulently attacked by Nasserists and other opponents.

The hostility between Saudi Arabia and Egypt which had been simmering in the late 1950s exploded after the creation of the Yemen Arab Republic in 1962 and the beginning of the Civil War:

2. *Newsweek* 21.12.1970

Egypt sent troops and weapons to the Yemeni Republicans and threatened to bring down the Saudi monarchy; and Faisal, in alliance with Hussein of Jordan, supported the Imam. Both leaders had their prestige at stake: Nasser could not afford to abandon the Yemen since his leadership of the Arab world had already taken a blow with the breakup of the United Arab Republic in 1961; and Faisal, with his position strengthened at home, was developing hopes of turning Saudi Arabia into a dominant force in the Arab and Islamic world, which was then dominated by nationalists: Nasser in Egypt and Qasim in Iraq.

The struggle for domination of the Arab world was on. While the Yemen War was the only military conflict and involved Soviet-supported Egypt against US-supported Saudi Arabia fighting for ideological domination on a broader field, the Yemenis fought their own war with greater or lesser help or hindrance from their respective allies. The war on the ground provided the backdrop for the diplomatic initatives which took place in the coming years. Nasser's tool was the Arab League which, founded in 1945 had, since he came to power become an instrument of Egyptian ascendancy in the Arab world. Through its auspices he set up the first two Arab Leaders Summits in Cairo and Alexandria in 1964, during which he discussed the Yemeni issue with Faisal. Given the trend in the Arab world in favour of progressive regimes, these summits were likely to adopt political positions influenced by Nasser which Faisal would regard as anti-Islam and pro-communist, so Faisal responded by trying rival summitry.

In December 1965 Faisal and the Shah of Iran launched the idea of an International Islamic Conference, and Faisal planned to visit a number of Islamic countries to promote the idea. After the fourth Arab summit in Cairo in 1966, relations between Egypt and Saudi Arabia deteriorated further as Faisal encouraged other regimes to object to domination of the Arab League by one state, which everyone knew to be Egypt. In August Faisal started a series of visits to Muslim states; by the end of the year he had obtained the support of Jordan, Morocco, Pakistan, Somalia and Iran. During this period the proposal for the Conference was violently attacked by Nasser, who compared it to the Baghdad Pact and described it as 'reactionary'. Before Faisal had the opportunity to make capital of his success and get the Islamic Conference off the ground, the June War broke out.

The June 1967 defeat reversed the balance of forces in the Arab world, thus making the Islamic Conference an idea of little relevance. The war marked the defeat of 'petty bourgeois' Arab nationalism and the strengthening of right wing regimes. This was not primarily due to the relative political strengths of the ideologies but rather to the relative financial power of the states involved. Egypt, the leader of the nationalists, was a poor, heavily populated country, which had already been bankrupted by the Yemeni war even before the June war and, by July 1967 was economically and politically in a shambles. In contrast, Saudi Arabia, the leader of Arab conservatism, had in recent years strengthened its financial position thanks to its increasing oil revenues, and had skilfully acquired political credibility. It was therefore, in the summer of 1967, in a position to call the tune in the Arab world.

The June War and after

The June 1967 defeat marked the end of nationalist domination in the Arab world. Although Nasser lived on and there appeared to be an increase in radical policies in the following years, the good days of Arab nationalism ended with the Israeli attack. Thereafter, Arab politics were dominated by Saudi Arabia and pro-US policies, and Saudi Arabia dominated the tone of negotiations. The Khartoum Summit Conference in August 1967 formalised the new relationship between Egypt and Saudi Arabia. In exchange for a regular subsidy, Nasser finally agreed to withdraw his troops from the Yemen. The 'Enemies of God' programme was stopped during the June war and never resumed, ending the publicity given to anti-regime activities and the encouragement to rebellion against the regime.

Although Saudi Arabia's domination over Egypt became clear after 1967, the apparent expansion of progressive movements after that date is often understood as a radicalisation brought about by the defeat of 'petty bourgeois nationalism'. Although left-wing movements did develop in the 1967-73 period, it is my suggestion that, despite appearances, Saudi domination of Arab politics was firmly established in the aftermath of the 1967 war, and that far from petty bourgeois nationalism being displaced in favour of truly 'progressive' movements, the bulk of the nationalist movement lost out to Arab reaction.

At first sight it appears that the Arab revolutionary movement expanded and developed after 1967. It was in 1967 that the

Palestine Liberation Organisation wrested its independence from the Arab League; in late 1967 that the People's Democratic Republic of Yemen achieved its independence under a radical leadership unrelated to Nasserism, and the *People's Front for the Liberation of Oman* became 'Marxist' and gained control of most of Dhofar. This period also marked the development of left-wing organisations in most Arab countries.

In my view, apart from the independence of the PDRY which, in 1969, took a radical turn to the left and has since developed as the only socialist state in the Arab world, these phenomena were of temporary significance and came about as a result of the 1967 defeat. The upturn taken by right wing reaction as a result of it was soon to put a stop to the radicalisation of the region which presented a threat to imperialism and to local conservatism.

The fact remains that the PDRY is the only regime which successfully developed against the prevailing trend. The *Palestine Liberation Organisation* never was truly independent of the Arab regimes, and, whenever it has tried to establish autonomy for itself, it has been crushed. This happened first in September 1970: the PLO refused to accept the American Rogers Plan for a solution to the Middle East question, which had been accepted by all the local leaders including Nasser. This refusal was made at a time when the movement's military strength was a threat to King Hussein's rule in Jordan. In September 1970 a war was fought in Jordan between the PLO and Jordanian forces which had outside support — the Palestinians were defeated with heavy losses and subsequently expelled from Jordan. More recently, after the 'victory' of the October 1973 war, and Kissinger's failure to obtain the desired settlement with Israel, the Saudi-dominated Arab states found it necessary to bring the PLO to heel in Lebanon, where the movement's strength had increased in recent years. Cutting the PLO down to size was one of the major aspects of the Lebanese Civil War in 1975-76. Throughout this period the PLO had been heavily financed by Saudi Arabia and other conservative Arab states despite the fact that some of the organisations within the PLO take strongly anti-Saudi positions. The bulk of the movement is aware of the limitations placed on it by the weight of the Arab states and Saudi Arabia in particular.

In September 1976, after the Lebanese Civil War had been raging for 18 months, it was on Saudi initiative that the Riyadh

Conference was held and an agreement forced on the conflicting parties. For months previously the Syrian President, Assad, had been successfully preventing any move which might stop the war before he had fulfilled all his objectives. Saudi Arabia was the only state able to force him to the negotiating table: within hours he accepted an agreement which did not give Syria all the power it wanted, but was on the whole favourable to it.

The left wing movement in Lebanon was not crushed by the Saudis, but certainly was one reason why the Civil War was encouraged to go on, as part of the concerted effort to smash the left. The Lebanese left and the Lebanese free press were the features of Lebanese society which gave most concern to the Saudis in the 1970s and the occupation of Lebanon by Syria, though it causes some concern, has certainly reassured the Saudis on this point.

In the same way that the PLO's struggle for independence within the Arab world failed, the struggle of the liberation movement in Oman, after a few years of expansion and success, was crushed by the intervention of Iranian troops in Oman thanks to the complicity of the Saudis, once conservatism had established hegemony over the region. It therefore appears that the rise of left-wing movements has failed as an attempt to oust Saudi Arabia either from its control over developments in the Middle East or as a leader of the conservative states.

Faisal's foreign policy between 1967 and 1973 achieved a strong measure of control over the confrontation states. While Nasser was still alive, Faisal encouraged him to accept the Rogers Plan and thus shift his position in favour of a US solution to the Israeli problem. Faisal also encouraged the War of Attrition which took place in 1969 and 1970. But it was only after Nasser's death and once Sadat was in control of Egyptian politics that the American line flourished. After Sadat's May 1971 victory over the left Nasserists, the road was open for an improvement in Egyptian-US relations and the easing out of the Soviets from Egypt. In June 1971 King Faisal paid his first official visit to post-Nasser Egypt on his way back from the USA. Saudi Arabia agreed to pay for Egyptian weapons if bought in the West, and the plans for a changeover in arms supplies were mooted, particularly after the expulsion of Soviet advisors in 1972. Even plans for an Arab arms industry, financed by Gulf states and located in Egypt have made some progress since then.

Saudi Arabian relations with Syria were improved after the coup by Hafez al Assad who, in late 1970, overthrew the more radical Jadid faction from power. In 1971, with Assad in power, economic overtures were made, transit facilities reopened and after a meeting in Kuwait between the Syrian Foreign Minister Abdel Halim Khaddam and the Saudi Foreign Minister Omar Saqqaf, Syria was included in the search for a US—oriented solution to the Palestinian problem.

Jordan and Egypt, and, after 1968, the PLO were financially dependent on Saudi Arabia, which tried to influence their policies towards a US-sponsored solution to Palestine. Saudi Arabia acted as an intermediary between the Arab and American leaders, and Faisal expressed his constant concern for the recovery of the occupied lands. This was, on his part, a genuine concern based on religion. As the leader of Islam, the guardian of the Holy Places, and a devoted believer, Faisal had stated on numerous occasions his desire to pray in Jerusalem, Islam's third holiest place, before he died. He wanted occupied lands returned to the Arabs, less out of solidarity with the Palestinians and concern for the liberation of Palestine, than from religious objections to Jewish presence at a site holy to Islam.

From 1972 onwards, while the Egyptians and Syrians were preparing in total secrecy for war, Faisal was the only Arab leader who was informed in advance of their plans. This was no mere courtesy to the paymaster, but was rather based on the reality that the limited military aims of the war could only be achieved if the struggle was accompanied by the power of Saudi Arabia through its possible use of the oil weapon on the one hand, and by influencing the USA on the other.

In the early 1970s, the Saudi Arabians had regularly denied any intention of using the 'oil weapon'. As late as November 1972 the Saudi Arabian Minister of Petroleum, Sheikh Yamani, explained Saudi Arabia's interpretation of oil power:

> '....we do not believe in the use of oil as a political weapon in a negative manner. We believe that the best way for the Arabs to employ their oil is as a basis of true cooperation with the West, notably with the United States. In this way very strong economic ties are established which will ultimately reflect on our political relations.'[3]

3. *Middle East Economic Survey* 3.11.1972.

Between November 1972 when this statement was made and May 1973 when the first suggestion of Saudi Arabia's use of the oil weapon was mentioned, a number of things had failed to happen. It had been Saudi Arabian policy in the previous years, particularly since Sadat had come to power, to persuade the Arab leaders that the solution to the Israeli problem could only come from the USA, since the USA was the only state which had power over Israel and could therefore compel it to return the occupied territories. The Saudis had succeeded in persuading the Arab leaders: the Egyptians had expelled the Soviet advisors, and prevented a progressive coup in the Sudan in July 1971, the Syrians had made their willingness clear after Assad's takeover, and the Jordanians had always been in agreement, while the Palestinians had been forced into a position where their objections could not prevent such a solution if it were agreed. But the Americans had made no response to the many Arab overtures, and the only signs of action were increasingly defeatist and conciliatory statements from the Arabs, while no comparable development was taking place in Israel or in the USA.

Hostility to this line of policy grew in the Arab world and, by early 1973, Saudi Arabia was losing its grip on Arab leaders. In March 1973 the Saudi Arabian embassy in Khartoum was seized by Black September guerrillas, leading to bloodshed, and revealing the extent of Saudi Arabia's loss of prestige: no Palestinian group would have dared make a military attack on the Saudis in any form or shape earlier, nor have they since. At that particular moment Saudi Arabia's prestige had sunk very low. The policies which the Saudis had forced onto everyone had failed.

Oil power and the October War

Faisal's awareness of his loss of prestige combined with his bitterness at the failure of his policies and, in May 1973, for the first time, he summoned the President of Aramco, Carl Jungers, and told him that 'Saudi Arabia was not able to stand alone much longer as the USA's only friend in the Middle East' and that he was subject to considerable pressure. The constituent companies of Aramco then immediately proceeded to take out full page newspaper advertisements in the American press calling for a more 'evenhanded' US policy towards the Middle East. The only concession obtained as a result, was the sale of Phantom jets to

Saudi Arabia, which had previously been refused on the grounds that the jets might be used to attack Israel. In June, the US, as usual, vetoed a Security Council Resolution deploring Israel's continued occupation of Arab territories.

Even once the October War had started, Faisal was reluctant to use the oil weapon despite the fact that Saudi Arabian officials had been making 'threatening' noises on the subject for months. It is also significant that Israel is the only serious problem on which Saudi Arabia and the United States disagree. In this context, Faisal's reluctance to take any action against the USA may be an indication of the country's dependence on the States.

Eventually under pressure from other oil producing Arab states, from a visit by a special emissary from Sadat, and particularly after the USA had snubbed the Arabs by voting $2.2 billion in emergency aid to Israel, the Saudi Arabians agreed to participate in a reduction of oil production and an embargo on the closest allies of Israel.

Although it did not influence any Israeli withdrawal in any way, nor bring about any change in US policies, the embargo worked like magic. Within hours Saudi Arabia had regained all the prestige lost in the last year and more. The oil price rise made unilaterally during the war and the subsequent one in January 1974 multiplied the financial resources of Saudi Arabia, giving the regime massive financial power over the Arab world and beyond. A well-managed public relations campaign in the West described Saudi Arabia as a heartless blackmailer, thus, at last, giving the mistaken impression that Saudi Arabia was on the side of the Arabs and not the prime supporter of the USA, an image which was very helpful to the Saudi regime in the Arab world.

Since the October War Saudi Arabia has been in a position of overwhelming strength in the Arab world, and has remained at the centre of negotiations over the Israeli problem. Saudi Arabian objectives in this field are therefore of some importance. Under Faisal the question of Palestine was understood primarily in religious terms and his main objective was to regain Jerusalem for Islam. As he himself put it:

> 'it is common knowledge that the Jews have no relation with Jerusalem, nor do they have any sanctuaries there... The Jews have no relationship with, nor right to exist in, and no authority or administration in Jerusalem...
>
> 'It is our duty, brothers, to move today, to move to save

our sanctuaries and drive out our enemies, and to be against all
the doctrines founded by the Zionists — the corrupt doctrines,
the atheist communist doctrines which seek to deny the existence
of God and to deviate from faith and from our religion of
Islam.'[4]

Apart from the question of Jerusalem, a priority for King Faisal,
the return of occupied Arab land, though not necessarily to the
Palestinians, was the main objective. Faisal was in no way
prepared to accept the continued existence of Israel on any part of
Palestinian territory, but his closeness to the US meant that he
was not prepared to support openly the total de-zionisation of
Palestine and the control of Palestine by the PLO. Faisal regarded
zionism as being the same thing as communism and both to be
cursed atheistic movements.

Foreign Policy under King Khaled

Palestine is one aspect of foreign policy which has changed
significantly since Faisal's death. We shall come to others below.
A mere two months after Faisal's death, King Khaled made a
speech in which, for the first time, the right of Israel to exist was
acknowledged by the Saudi regime:

> 'We are doing our duties vis-à-vis Arab brothers and will
> also fulfil our obligations towards the Palestinian state once
> it is established. At such time when these things have been
> accomplished and Israel has withdrawn from all the
> occupied territories, including Jerusalem, it can live within its
> 1967 borders.'[5]

King Khaled also promised the future Palestinian state on the
West Bank considerable financial aid, and again revealing Saudi
influence on the Arab world, gave the US some advice on arms
sales and anti-communism:

> 'We wish the United States of America would decide to
> arm Egypt and Syria. Then billions would flow into the
> American Treasury', he said. If the United States pressed
> Israel into making a just peace settlement, 'Russia would not
> acquire a single foothold in the Middle East.'[6]

While Faisal's role in Arab politics was usually behind the scenes,

---

4. BBC *Summary of World Broadcasts*, part 4, 1.1.1974.
5. *International Herald Tribune* 26.5.1975
6. *ibid.*

King Khaled has participated far more publicly in the search for a solution to the Palestinian problem. His greater flexibility and sensitivity to world politics have resulted in the statement above explicitly making the conditions under which Saudi Arabia is prepared to accept an Israeli state in the Middle East.

Khaled's participation in the conflicts of the Arab world since 1975 is particularly clear in the case of the Lebanese Civil War. The war was allowed to rage on until it appeared that the outcome would be counterproductive in terms of Saudi Arabian objectives, which were the control of the Palestinian movement, i.e. the reduction of its autonomy so as to compel the movement to accept a solution agreeable to the Arab states. Saudi Arabia also wanted to use the war to smash the left in Lebanon, which in the 1970s had grown to be an important force in the Arab world thanks to its widely distributed press and publications, to ensure Lebanon's future as a service and trade centre, and to prevent the Syrians from achieving their resurrected 'Fertile Crescent' plan with Jordan, Lebanon and a West Bank Palestinian state, all under Syrian domination. While the Palestinians were still strong, the Syrians were allowed to pursue their alliance with the Right Wing Lebanese Christians in an attempt to annihilate them. When the Syrians seriously started to consider reviving the Fertile Crescent idea with the entire area under their influence, pressure was finally brought to bear on the conflicting parties, bringing them to the conference table in Riyadh. Thanks to Saudi Arabian financing, the Syrian invasion force was transformed overnight into an 'Arab League Peace Keeping Force', with a few Saudis and other Arab soldiers added for decoration. Since then, however, the Syrians have been prevented from overstepping the mark.

In 1976 and 1977 Saudi Arabia and the other Arab states have made every possible move to ensure some kind of settlement. Even the PLO has shown its willingness for a settlement acceptable to the Arab states. The new administration in the USA even came close to resorting to the old Kissinger shuttle, with their Secretary of State Cyrus Vance, and in early 1977 the world was led to believe that the Israeli general election would be followed by pressure from the USA, a Geneva conference in the autumn, and at long last — a *solution*.

Two factors contradict this fairy tale: first the Likud won the Israeli elections and indicated willingness to negotiate, claiming 'everything is negotiable' but that the West Bank is 'liberated', not occupied. Since its victory the Likud has been busy legalising the illegal Jewish settlements on the West Bank, and planning more of them. The other factor is that, despite this shocking setback to President Carter's policies, the US government has shown no inclination to put even the mildest form of pressure on the new Israeli government composed of former terrorists of the British Mandate period in Palestine. On the contrary, the US administration has stated clearly that pressure would not be put on Israel. In this context the prospects for peace in the Middle East looked bleak at the end of 1977.

Foreign policy since 1975

Saudi Arabia became a dominant power in the Arab world after 1967, but it was only after the massive oil price rises in 1973 and 1974 that the country reached a position of importance on the world scene thanks to its financial clout.

Although certain fundamental features of Saudi foreign policy have been carried over from the earlier period it is only since 1975 that the state has been able to expand and implement them. Since King Khaled's accession, a new feature of Saudi Arabian foreign policy has been its flexibility, which can be seen in the way relations with the Gulf and Red Sea states have been handled. Another comparatively new factor has been the disposal of surplus oil revenues by investment, largely in the West, thus giving Saudi Arabia a new measure of influence in the advanced capitalist countries. The other objectives have been carried over from the Faisal period: they are the propagation of Islam and the consequent cooperation with Islamic states as well as efforts to influence them towards greater respect for Islam. Anti-communism has remained a feature of Saudi Arabian foreign policy and the fact that it is now practiced with greater flexibility does not imply a reduction of the regime's hostility to communism. We will in the following pages look at these aims and they way they have affected Saudi Arabia's foreign policy in certain specific areas.

The Peninsula

The change in style of Saudi Arabia's foreign policy is nowhere more obvious than in its relations with the People's Democratic Republic of Yemen. It was regarded with hostility at independence in 1967, and by the time the regime shifted to the left in 1969, Saudi Arabia was already supporting right wing exiles residing both in the Yemen Arab Republic and in Saudi Arabia itself. In November 1969 Saudi Arabian troops were sent to attack the PDRY's border post at al Wadiyah and were backed by the Saudi Air Force. Although they won on the ground, they did not pursue their advantage and in the following years Saudi Arabia has merely continued to support exile groups giving them supplies and facilities such as the Radio of the Free South.

In 1972 the Saudi regime used its close relationship with the ruling group in the Yemen Arab Republic to encourage a full-scale war between the two Yemens. This broke out in October 1972 and, contrary to expectation, the Saudi-supported North Yemenis and exiles were unable to make any headway against the PDRY's popular resistance. Various Arab states offered to mediate, and a ceasefire was soon arranged in Cairo. Its announcement marked a serious setback for Saudi Arabian policy against the PDRY as the two Yemens announced their decision to unite into a single state within a year. This had not happened by 1977, although in the intervening years, unity committees of political and economic policy have continued to meet regularly at various levels, and relations between the two states with opposite political positions have been far better than could have been predicted. Another Saudi Arabian plan under Faisal was to invade the Hadhramaut (part of the east of the PDRY) and separate the PDRY from its eastern region. From Faisal's point of view this would have been beneficial in two ways: first the Yemenis would no longer be able to suport the nationalist struggle against the Sultanate of Oman, and second, Saudi Arabia would obtain an outlet on the Indian Ocean. This plan was not implemented, but intense hostility continued.

Since Faisal's death Saudi Arabian policy towards the PDRY has taken a radical turn. Within months of Khaled's accession to power there were rumours of Saudi mediation between the PDRY and the Iranian occupation forces in Oman. In March 1976 diplomatic relations between the two countries were established, only a few days before King Khaled's tour of the Gulf States. It is clear that Saudi Arabia hoped to moderate the policies of the PDRY's government by using the carrot rather than the stick, and to persuade the PDRY to reduce its aid to the *People's Front for the Liberation of Oman*. Since recognition, there has been talk of a number of plans for economic cooperation between the two states. In particular a pipeline has been planned running from the main Saudi Arabian oil fields to the Aden refinery or to a new oil port in the Hadhramaut. The idea behind these plans is to give the Saudi Arabians greater independence from the Gulf Straits as an outlet for their oil. It would also give the impoverished PDRY a most welcome addition to its national income. There has also been talk of financial aid to the PDRY, but as of mid-1977 this remains unconfirmed: the sum of $50 million has been mentioned.[7]

Although by 1977 there seemed to be little change in the political options of the PDRY there had been some softening on foreign policy issues. In December 1976, during a visit to Paris, the PDRY's Foreign Minister declared his country's willingness to recognise the Sultanate of Oman should Iranian troops be withdrawn. Whether the partial withdrawal which took place in early 1977 will lead to recognition is doubtful, since the base for Phantoms which threatens the PDRY was one of the aspects of Iranian presence in Oman which was not changed in 1977. In the course of that year, however, further 'secret' meetings were reported to have taken place between PDRY and Omani officials under Saudi Arabian auspices.

In the Red Sea area, an indication of Saudi Arabia's new flexibility can be seen over relations with the Horn of Africa. While, under Faisal Saudi Arabia had opposed the Eritrean Liberation movement as a whole, which was described as 'communist', the new regime has given considerable aid to some sections of the Eritrean movement. It has supported the movement on the grounds that it is Arab and it has given considerable aid to strengthen the relative position of 'moderate' organisations within the Eritrean movement at the expense of the avowedly Marxist organisation. Similarly Saudi Arabia has tried, and to some extent succeeded, in wresting Somalia away from Soviet influence, by offering to pay for Somali arms purchased in the west, and by bringing Somalia into the Arab fold. Relations with the Yemen Arab Republic have continued to be excellent and that regime is still extremely dependent on Saudi Arabia for financial support, despite some political differences due to the complex social and political structure of the YAR. The YAR receives a grant of SR345 million a year from Saudi Arabia.

The new regime has also made significant changes in the

7. *Le Monde* 27.1.1977

relationship of Saudi Arabia with other Gulf states. These relations continue to be dictated by a number of factors: (a) Saudi-Iranian rivalry for domination of the lower Gulf states; (b) hostility towards progressive movements which are believed to be influenced by communism; (c) exclusion of any Soviet influence in the area.

Rivalry between Iran and Saudi Arabia to wield influence over the lower Gulf region is an important factor in the relationship of the two states. Fear of Iranian domination is shared by all the gulf rulers except Sultan Qabus of Oman who invited the Shah of Iran to assist him to defeat the People's Front for the Liberation of Oman. Qabus has therefore brought Iranian forces to the Arab side of the Gulf, arousing concern among the other states. Saudi Arabia, although it disapproved of this presence, was militarily too weak to oppose it. All the conservative regimes in the Gulf agreed that the PLO had to be defeated, but the Arab regimes objected to Iran's military intervention as it posed a long term threat to themselves, given the considerable Iranian population in their own states and the Shah's well-known ambition of controlling the entire region. Despite this, none of them, including Saudi Arabia, opposed the Iranians openly, as they all recognised that Iran was the only state with the military power to crush the Front. But its intervention caused unease, and when, in 1975 it became clear that the Front had effectively lost control of Dhofar, the Saudi Arabians were the first to suggest that the time had come for the Iranians to go: Prince Fahd stated that 'Saudi Arabia opposes any outside interference by any party in the Sultanate of Oman. The Sultanate must be left to manage its own affairs without outside interference.'[8]

In March 1976, King Khaled made a tour of Gulf states shortly after recognising the PDRY, and during his trip made a number of statements concerning the situation in Oman and suggesting that since the Dhofar issue was settled, diplomatic relations should be established between the PDRY and Oman.

The situation in Oman has been at the centre of Saudi-Iranian rivalry in recent years, but it is not the only subject of disagreement between the two states. Another, related, issue has been the much discussed Gulf Security Pact which Iran has been

8. *Al Rai' al 'Am* 20.11.1975

promoting, and which would ensure Iran's domination of the region. As early as May 1975 King Khaled expressed his objection to such a Pact, saying that 'Cooperation in all fields among the Gulf countries can be achieved even without the existence of alliances'.[9] In November 1976 the first Conference of Gulf States Foreign Ministers met in Muscat and the conference was expected to formalise the alliance. But while it was going on, PDRY ground forces shot down an Iranian Phantom jet, causing an outcry as it was revealed that the Iranians had been flying missions over the PDRY regularly. The meeting broke up in disarray and it seems unlikely that such a pact will be formally established.

Under Faisal, relations with the lower Gulf states were tense. Historic border disputes with Abu Dhabi marred relations, while the House of Saud got on well with its sister Wahhabi state, Qatar. The formation of the United Arab Emirates in 1971 and the independence of Qatar and Bahrain earlier that year revived the Saud's dormant ambition to dominate the entire peninsula. The regime only recognised the United Arab Emirates in August 1974 after a border agreement with Abu Dhabi was signed in July of that year, finally settling the Buraimi dispute; the agreement gave Abu Dhabi some concessions but ensured Saudi Arabia a route to the Gulf coast for a pipeline. Despite this reconciliation, ambassadors were only exchanged in June 1975 after Faisal's death, and it was only in March 1976 that King Khaled made a tour of the region to strengthen relations with the regimes. Most of these statelets present no problems to the Saudis as they are as reactionary and undemocratic as Saudi Arabia. Kuwait and Bahrain are slightly different cases.

In both Kuwait and Bahrain, some form of democratic consultation existed. In Bahrain, this had only been since 1973 and in Kuwait since 1961. Close relations with Saudi Arabia were nevertheless maintained, but it is clear that the liberal attitude implied by democratic institutions was not welcome to the Saudi regime which feared that democracy might have a tendency to spread. In August 1975 the Bahraini parliament was closed down after only two years of operation, because it proved less than compliant with the ruling family. Since then plans for a causeway

9. *International Herald Tribune* 26.5.1975

between Bahrain and the Saudi mainland which had made little progress earlier have reached the stage of detailed planning. Despite fear of cultural influence which is experienced by both regimes concerning the other, relations are excellent and the Bahraini refinery is dependent on Saudi oil for continuing operations, and Saudi Arabia even 'gave' Bahrain half an oil field.[10]

The case of Kuwait is different insofar as Kuwait has no need for financial donations from Saudi Arabia, but the existence of a rather liberal parliament there as well as of a press which, once the Beiruti press was silenced, was the only free press remaining in the Arab world, created some tension. In August 1976 the Kuwaiti parliament was dismissed and strict controls put on the press. Although this was at first described as temporary, by mid-1977 it appeared unlikely that Parliament would reopen in the foreseeable future. This outcome to the only democratic experiments in the Gulf could only please Saudi Arabia in whose interest it is to prove that 'Islam is the best constitution in the world'. As both these experiments appeared to fail for internal reasons Saudi Arabia could only take credit for its own political line!

The third world

In its relations with the third world, beyond the Arab lands, Saudi Arabia considers assistance to Muslim states to be a priority over aid to other countries. Saudi Arabia is a leading member of the Jeddah-based Islamic Development Bank which makes loans in *islamic dinars* (1 Islamic dinar = Special Drawing Right 1). Pakistan is the country which has benefitted most from Saudi Arabia's Islamic solidarity. Relations between the two countries have always been close, and Pakistan can rely on substantial financial aid from Saudi Arabia. For example when King Khaled visited Pakistan in October 1976 he left behind a donation of $30 million. But despite statements to the contrary Saudi Arabia does interfere in other states' internal affairs. As we have seen, the Saudis have hardly kept out of Arab governments' internal affairs, and in Pakistan Saudi support for President Bhutto's opponents in the conflict which took place after the election crisis in 1977, certainly contributed to the President's downfall. This happened despite the fact that Saudi Arabia was supposedly

10. *Financial Times* 1.11.1976

acting as a 'mediator' between the two factions involved in the dispute, and its support for Bhutto's opponents was due to their political positions which were more conservative and Islamic than Bhutto's: soon after the military takeover in the summer of 1977, the government announced that in future Islamic law would be implemented in Pakistan.

Saudi Arabia has tried to persuade other states to re-adopt Islamic law which involves the use of barbaric punishments such as the amputation of the hand for theft. Pakistan is not the only state to submit to these pressures though, there are plans to reintroduce Islamic law in Egypt and the Sudan.

Saudi Arabia has also supported the Muslim autonomy movement in the Philippines and has tried to persuade President Marcos, otherwise regarded as a reliable reactionary ally, to come to terms with the movement whose religious aims are approved of by the Saudis.

Beyond the Arab world Saudi Arabia's influence on the Third world is exerted through financial aid projects. Saudi Arabia has substantial capital in the Arab Bank for Economic Development in Africa, and the Arab Investment Corporation. Most Saudi financial aid goes to poorer Arab states, such as Tunisia, including a SR105 million loan for a sewerage system, Morocco

Foreign aid committed by Saudi Arabia ($million)

	Bilateral Grants	Loans	International & regional institutions	Total aid	Aid as %GDP
1972	143.1	77.8	-	220.9	2.7
1973	650.2	343.6	-	993.8	5.0
1974	1,130.6	747.6	2,203.3	4,081.5	12.4
1975	1,363.5	2,037.6	2,262.9	5,664.0	13.8
Total	**3,287.4**	**3,206.6**	**4,466.2**	**10,960.2**	**10.6**

Source: Ministry of Finance and National Economy

Mauretania and Bangladesh. In countries where there is a substantial Muslim population Saudi Arabia pays for the building of mosques and other facilities; in Uganda, for example, Saudi Arabia's contributions to agricultural projects is due to the presence of a Muslim minority in the country. Saudi Arabia has financed an aluminium scheme in Guinea and paid for roads and ports in South Korea. South Korea also benefits from Saudi assistance in the form of substantial building contracts as part of the Saudi Development Plan, as do some other anti-communist third world states such as Taiwan.

The special relationship with the USA

Since 1975 Saudi Arabia has been able to increase its influence in the world largely thanks to its oil revenue surpluses for which the centres of capitalism have been competing to have invested in their countries in order to prop up their own weakened economies. Saudi Arabia's links with the USA are particularly strong in this field.

Although there are political disagreements between Saudi Arabia and the USA, particularly over Israel, the USA is undeniably Saudi Arabia's closest ally and there is a strong element of mutual dependence in the relationship. The USA is increasingly dependent on Saudi Arabian oil and Saudi Arabian goodwill in the search for a solution to the Palestinian problem, as well as on the investment of Saudi Arabian surplus cash in the States. Saudi Arabia is dependent on the USA for military training and supplies, for the importation of technology for its development plans and for educational and other professional services.

After the quadrupling of oil prices in 1974 the USA rapidly worked out that the fastest way to 'recycle' petrodollars was to sell military hardware to the oil producers. From $176 million in the twenty year period between 1950 and 1970, total government to government military sales from the USA to Saudi Arabia shot up to $2.510 billion between 1971 and 1975. In the last four months of 1975 the US Congress was asked to approve sales of a further $3.5 billion, including $1.4 billion to build a military city at al Batin, only 40 miles from the Iraqi border, and $158 million for a National Guard headquarters.

In 1975 a controversial contract was signed for the training of the Saudi Arabian National Guard by employees of a private US firm, Vinnell Corporation. The contract 'to supply military advisors to form and train four mechanised infantry battalions and one artillery battalion for the National Guard'[11] was signed

---

11. *Washington Post* 31.7.1976

by the US Defence Department on behalf of the Saudi Arabian Government and was officially worth $77 million, but believed to be well over $100 million. Furthermore it was signed only a few weeks after Kissinger had threatened to occupy Saudi Arabian oil fields should there be another embargo, which raised the question of the value of having one's military forces trained by the very people who are threatening to invade one's country. A former Army Special Forces officer partly answered this question pointing out that all the men he knew on the Vinnell payroll were ex-US Green Berets and other Special Services men, and that he had evidence linking the organisation with the CIA. He added:

> 'Our guys aren't over there training the Saudis against what might someday be an American invasion of oil fields. That's a super-con, a political feint. Our guys are over there to supply and train the Persian Gulf Armies to cope with their own security problems.'[12]

Other aspects of US-Saudi military cooperation have aroused protests, particularly the enormous bribes paid by US arms manufacturers to various intermediaries to obtain highly profitable sales contracts. These concerned mainly Northrop and Lockheed, the latter having paid $106 million over five years to Adnan Khashoggi alone!

Saudi Arabia is the only country in the world outside the USA where major building projects are managed and run by the US Army Corps of Engineers who have been involved in Saudi Arabia since the 1950s and have designed and supervised the building of the Dhahran airfield, Saudi Arabia's television network and various military installations. In 1976 they were involved in the construction of naval facilities, military cantonments and housing, and at al Batin in the construction of a $3 billion city for over 50,000 people, including headquarters, hospital and all facilities.

In late 1975 there were altogether over 3,000 US citizens involved in Saudi military affairs. If one remembers that the Saudi Armed Forces only total about 70,000 men, including the National Guard, the figure is not insignificant. Military sales also shot up from $1.4 billion in 1975 to a proposed $4.5 billion in 1976. The US also gave Saudi Arabia assistance by training police officers both in the US and in Saudi Arabia. The Agency for International

---

12. *Village Voice* 24.3.1975

Development's Office of Public Safety directed a $385,000 study of police communications, criminal identification, and other aspects of internal security for the Saudi Arabian government in early 1975 just before it was abolished when Congress defined it as 'a programme for the preservation of right wing dictatorships throughout Latin America and Asia'.[13]

At the directly financial level, Saudi Arabia's current account surplus is estimated to run at $20 billion annually and its foreign assets to total some $50 billion in mid-1977. Although details are hard to come by, a significant portion of this surplus is invested in the USA in government paper, such as Treasury bills, bonds and notes. Some 70 per cent of Saudi assets are in dollar-denominated instruments in the US and Europe.[14] The speed with which foreign assets are increasing is likely to turn Saudi Arabia into a capital exporting country of major proportions by the early 1980s, and thus enable it to invest as it chooses throughout the capitalist world. Saudi Arabia's foreign holdings have risen by $44,803 million between early 1974 and the end of 1976.

US-Saudi technical involvement is such that even the Saudi Development Plans have been drawn up with American assistance, but American firms, ranging from the Corps of Engineers, to the Bechtel Corporation, the Hospital Corporation of America, Wilson Murrow and hundreds of others, are heavily involved in all sectors of the Saudi Arabian economy.

The link between Saudi Arabia and the USA is, as we have seen, one of mutual dependence. Another aspect of their good relations is the fact that their international objectives coincide. Both states oppose communism and want to promote and support 'free enterprise' on the American model throughout the world. Although it is often said that the Saudi government is a 'US puppet' this is in fact incorrect: the reality is that the interests of both states converge on most points. The US want to retain their domination of the world, and specifically access to Gulf oil, which they will increasingly need in the coming years. Both states want to ensure the survival and continuation of conservative, and hopefully strong Islamic regimes in the Middle East and prevent the development of any 'communist' influence in the region.

The Saudi regime needs the USA for its development and also

---

13. *Washington Post* 2.11.1975
14. *Middle East Economic Survey,* 29.8.1977

for its protection against rival states in the region or against internal dissent. Its oil reserves on the one hand and its foreign financial holdings on the other mean that Saudi Arabia's power on the world scene will increase in the coming years, and although it was true to say in the past that Saudi Arabia needed the States more than the US needed Saudi Arabia, this is no longer the case. Even the balance of trade between the two countries has recently been reversed, and Saudi Arabia now exports more to the US than the US to Saudi Arabia which, given the amount of Saudi imports, is remarkable.

	US exports to SA ($millions)	US imports from USA	Est. balance
1975	1,501.8	2,623.3	(−1,121.5)
1976	2,774.1	5,213.4	(−2,439.3)

Saudi Arabia's support for the 'free world' is not limited to the USA. In recent years the state has invested some of its surpluses in aiding such ailing capitalist economies as the Italian and the British. Saudi Arabia has a number of military aid contracts with Britain, including a £500 million contract for the development of the Royal Saudi Air Force signed in 1977. Estimated to be worth over £850 million when inflation and additional requirements are included, it is the biggest export contract Britain has ever obtained.

Part of Saudi Arabia's attempt to reduce dependence on the USA has taken the form of increased bilateral relations with European countries and Japan. Agreements have been reached for the sale of oil to their governments and for these countries' assistance with the implementation of the Development Plan. Such agreements have involved infrastructure projects, consultancy, participation in oil-based projects, as well as military sales and assistance with education and health. By 1977 it was true to say that although the Americans still held the lion's share of contracts in Saudi Arabia, other western states were competing well.

The communist world

Saudi Arabia's virulent anti-communism makes it easy to forget that the Soviet Union was the first country to recognise Abdel Aziz as King of the Hejaz in 1926. As a result Abdel Aziz sent his son Faisal, the future King, to visit the Soviet Union in 1932, at a time when the USSR was giving Saudi Arabia economic assistance in the form of cheap fuel and food. After the oil concession was granted to SOCAL, relations were broken off in 1938 and since the foundation of Israel, communism has been regarded, particularly by King Faisal, as an arm of the Jewish world conspiracy. At times of stress in the relationship with the West, there have been rumours of plans to establish diplomatic relations with the Soviet Union and to negotiate trade agreements, as in the mid-1950s, but none of these plans has ever materialised.

Throughout the post-war period, hostility has been mutual, the Soviet Union describing Saudi Arabia as a 'theocratic, feudal-absolutist state' whose structure was 'arch-reactionary class essence',[15] while to Saudi Arabia 'the role of the Soviet presence on Arab land as represented in military and economic agreements has begun to reveal methods of subversion and various forms of pressure in favour of Zionism, the first and last mother of communism.'[16]

Despite these mutual insults, in recent years it has been the custom for both states to exchange greetings on their respective National days. When it has suited either of their international policies hostile statements have been milder and rumours of a re-establishment of relations current. In 1964 when Faisal finally became King, he waived the usual visa regulations to invite a Soviet journalist to visit Saudi Arabia and the journalist wrote some sympathetic reports. In the 1960s despite the lack of diplomatic relations, trade developed between the two states.

Relations have not changed significantly since Faisal's death although there appears to be a softening of Saudi policies and, as part of the liberalisation of foreign policy, hints of improved relations with the Soviet bloc have been made. In May 1975, King Khaled indicated that when he said: 'We are not against the Communist states *per se*, but against communism as a dogma and, therefore, we will protect ourselves against it by all means.'[17]

Saudi Arabia has established no diplomatic relations with any communist state and is unconcerned with Asian communist states insofar as they do not figure in its chief international pre-occupation: the Palestinian problem. Saudi Arabia does however

15. Stephen Page, *the USSR and Arabia 1955-1970* p.80
16. BBC *Summary of World Broadcasts* part 4, 18.8.73
17. *International Herald Tribune* 26.5.1975

support anti-communist regimes throughout the world, in Latin America and South Korea and Taiwan in Asia among others.

Conclusion

Within the space of 40 years Saudi Arabia has leapt from being a totally insignificant state to one whose place in the world cannot be ignored. A steadfast supporter of conservative regimes, the country has, since 1974, increased its financial holdings to such an extent that it is in a position to qualify for its own seat on the board of the International Monetary Fund. According to one prediction, 'by 1981 Saudi Arabia's income from its assets may well nearly equal its savings from oil production'.[18]

This means that not only will Saudi Arabia be a financial power in the third world and in a position to determine economic policies there, but it will also export capital to the industrial countries who will thus be dependent on it. Given the regime's political options, it is unlikely that it will choose to use this power in an attack on capitalism, and the West will make sure of that.

At the Arab level, Saudi Arabia's domination of the Arab world, and particularly of the non-oil producing states, is a fact which is likely to continue for decades to come. Egypt, for example, is totally dependent on Saudi Arabia in all its economic policies. Despite the fact that the Saudis are not happy with the lack of development and economic reforms there, when the Egyptian people rioted in the streets for bread in early 1977, Saudi Arabia and other oil producers forgot about the conditions of their aid to Egypt. It is unlikely that a solution to the Israeli problem could be implemented without Saudi Arabian support for it, and that means if there is to be a solution, it will have to include the return of Jerusalem.

CHAPTER VII
THE ECONOMY
AND
DEVELOPMENT PLANS

Before the Development Plans there had been no attempt to create any coherent programme for economic development. With oil revenues supplying the means to pay for imports, little thought was given to investment in agriculture or industry to generate a surplus within the country and reduce dependence on imports. Nor was there much effort to develop human resources to reduce the dependence on migrant and contract labour.

Those developments which had taken place before the 1960s were almost entirely off-shoots of the oil industry — both in financing and in the elements of infrastructure which were laid. The US Bechtel Corporation built for Aramco everything required ranging from family housing, pipelines, marine terminals, power plants, and central air conditioning systems, up to a value of half a billion dollars between 1944 and 1957, and engaged in construction projects for the government — completing a power plant for the electrification of Riyadh in 1949, building roads in and around the capital, and re-laying the Riyadh airport to provide adequate landing facilities and repair shops, warehouses, offices and living quarters. In Jeddah the company built administrative buildings, hangars and workshops for the airport (completed in 1949) and opened a new power plant; it also built a deep water pier to cope with the problem of unloading ships which were arriving in increasing numbers with the steep rise in imports. To build this pier Bechtel demolished the historic city wall and used the stones to fill the pier which was completed in 1950. Bechtel was also responsible for the Dammam to Riyadh railway, a single track line whose services were frequently interrupted because the track was covered in sand. The railway was financed partly by a $15 million loan from the US Import-Export Bank and the rest ($52.5 million) from Aramco — the Saudi government was to repay this in installments by 1960. Running costs of the railway have proved exorbitant, but it is still in use.

Apart from oil drilling itself, and the construction of roads and ports to facilitate imports and royal travel there was some development of local private enterprise in the Eastern region, but this again was largely to service Aramco and Aramco workers. In 1956 the company set up the Arab Industrial Development Department to encourage local industry through loans and training, and thereby to cut down on the need to import all its requirements. Vehicle repair shops, poultry farms, transport firms leasing out lorries, and construction related industries were among the local enterprises developed in the 1950s.

The impact of the oil economy, dependence on ever-increasing imports placed pressure on airports, seaports and roads, and the creation of new Ministries all stimulated a boom in construction. To supply the construction industry, there were by 1965 two cement plants in Jeddah and Hofuf, supplying 600 tons a day, and another under construction in Riyadh. Other factories produced cement pipes, bricks and, in Riyadh, a gypsum factory was producing 100 tons of wall panels a day.

Petromin (see above p.53) was set up in 1962 to develop an oil-based industry, and the Jeddah steel rolling mill still has a potential for providing a base for manufacturing industry. The planned capacity of the mill was 45,000 tons a year, but by 1970 it produced a mere 8,500 tons. Badly planned and constructed there was no adequate staff to run it and it therefore operated only one shift per day instead of three. Despite this initial failure, the mill is regarded by the regime as the first stage of an integrated minerals industry in the country, where considerable iron deposits are believed to exist.

There was also some growth in the local food industry, again centred mainly on the eastern region. By the mid-1960s there were three modern dairies, two date picking plants and a large number of soft drinks bottling plants, as well as ice-making factories and cold storage depots.

Before the Plans, government efforts to develop local industry were sporadic. In 1963 the Protection and Encouragement of National Industries Ordinance was promulgated to assist local entrepreneurs to set up import-substitution industries and compete with imports. They were offered financial assistance, such as the right to import machinery and raw materials without customs duties, and free land for their factories and living quarters for their employees, as well as an increase of tariffs on the import of goods competing with local industry. Although the Ordinance did not have an enormous influence in persuading reluctant traders and middlemen to diversify their interests into productive industry, by the end of 1971, 188 firms had been licensed. Their activities however do not appear to provide a basis for the development of local self-sufficient production or the creation of a surplus nationally. Of the 188 firms fifty were in construction and a number of others in related industries — furniture making, metal pipes and frame fitting. The rest were mainly in food, clothing and printing. All of them depended on imported materials.

By 1970 industry was still at an embryonic stage. Of 8,174 manufacturing establishments in 1967-68, 89 per cent employed between one and four people, i.e. they were workshops of the traditional variety — bakers, tailors, carpenters, etc. The rest were divided between 'import substitution' industries like drinks bottling and the assembly of consumer goods, none of which in fact substitute for imports since all the materials are imported, and construction related industries, which rely on a continual boom in building.

In the late 1960s three industrial estates were being planned in Jeddah, Riyadh and the East. The locations chosen indicate the intention to develop further the areas which were already the main centres, where economic expansion has already taken place — the East because of oil, Riyadh the royal and administrative centre, and Jeddah the sea port where the Hejazi bourgeoisie were beginning to diversify from trade into industry. In all these centres population has grown and with it both the demand for consumer goods and the supply of labour.

THE FIRST DEVELOPMENT PLAN 1970-75

The First Development Plan was the first attempt by the Government to provide a comprehensive approach to economic development. As early as 1961 the Supreme Planning Board had been set up, but nothing was done, and after years of chaos, the Ford Foundation was invited to recommend improvements. As a result the organisation was renamed the Central Planning Organisation, and was responsible for the plan in cooperation with the Stanford Research Institute of the USA, which was hired for the purpose in 1968.

The Plan can best be seen as an attempt to rationalise and modernise economic development, alongside the attempts in the 1960s to modernise the administration. An immediate incentive behind its creation was the deficit in the budget brought about by Saudi Arabia's financial commitments to the Confrontation States after the June 1967 War. Although, except for the year 1968-69, national income continued to rise, after 1967 it failed to keep up with increased expenditure, as can be seen in the following table:

Table 1

Government Income and Expenditure 1966/7-1969/70
(SR millions)

	1966/7	1967/8	1968/9	1969/70
Revenues	4,823	5,137	4,957	5,668
Expenditure	4,570	5,374	5,602	6,079
current	3,782	3,040	3,080	3,254
development	788	1,704	1,860	2,163
aid to Arab countries	—	630	662	662
Surplus/deficit	+253	−237	−645	−411

Source: SAMA

As well as being planned in a period of financial stringency, the Plan could not be expected to foresee the historic changes which were to take place in the world oil trade during the Plan period. Nor therefore could the planners have predicted the extraordinary increase in revenue and development potential which was to take place after 1973.

In the Plan SR41.3 billion was allocated for expenditure, of which SR18.4 (44.5 per cent) was to go into capital projects.

Table 2 Summary of five-year plan allocations
(million rials)

	Recurrent	Project	Total	%
Administration	6,794.6	922.8	7,717.4	18.6
Defence	3,980.0	5,575.0	9,555.0	23.1
Education and training	6,150.2	1,227.5	7,337.7	17.8
Health and social affairs	1,612.9	308.2	1,921.1	4.7
Public utilities and urban development	1,246.9	3,325.4	4,572.3	11.1
Transport and communications	1,767.3	5,709.2	7,476.5	18.1
Industry	321.8	776.7	1,098.5	2.7
Agriculture	973.8	493.9	1,467.7	3.6
Trade and services	83.5	43.8	127.3	0.3
Total	22,931.0	18,382.5	41.313.5	100.0

Source: Central Planning Organisation: Development Plan 1390 (1970)

The Plan was not actually fulfilled as, although 72 per cent of the planned expenditure was allocated during the first three years, only 60 per cent of these allocations had been spent by the end of the third year. Overall the Plan aimed at increasing GDP by 10 per cent a year, and this was fulfilled easily. This however does not give a real indication of the development of the economy, as the increase in GDP is almost entirely due to the multiplication of oil revenues after 1973, rather than to increased output in industry or agriculture. As we shall see, agriculture failed to reach its target. Although GDP measurements can be useful in certain cases, in countries like Saudi Arabia, they are totally distorted by the presence of a single factor, in this case oil, which gives a false impression of economic growth; the extraction and sale of crude oil cannot in itself constitute economic development unless associated with the development of a productive national economy.

Given that the Plan's stated objectives were to increase economic growth, develop human resources and diversify the economy to reduce dependence on oil, the proportions devoted to industry and agriculture are an indication of the seriousness of the commitment to these aims — agriculture was allocated 3.6 per cent and industry 2.7 per cent. Although allocations to health and education amounted to 22.5 per cent of the total budget, this may not be such an impressive ratio in a country where the literacy rate was estimated at between 10 and 15 per cent in 1971-72, particularly since investment was concentrated on buildings rather than on the training of teachers. Education and health put together had a lower investment than defence, which received 23.1 per cent of allocations.

In agriculture the Plan aimed for a 4.6 per cent increase in production yearly over the Plan period: the real rate of growth turned out to be 3.6 per cent and even the Second Development Plan recognises that these figures 'are based on national accounts data, which may overstate the actual rate of growth'.[1] In 1974-75 agriculture accounted for only 8.6 per cent of the *non-oil* gross domestic product. During the Plan period the agricultural labour force diminished by almost 1 per cent a year as a result of increased poverty of farmers and their attraction towards the towns where they migrated to obtain employment, usually as unskilled labourers in the boom sectors such as construction. At the same time the rate of food consumption increased, largely as a result of the growing migrant population at all levels of the labour force. Food imports grew by over 13 per cent during period.

The depressing results in agriculture might lead one to hope that in the industrial field progress might have been greater, but it wasn't. Even within the oil-related sector, the rate of growth was well below that planned: oil refining grew by 4.1 per cent compared to the planned 9.1 per cent: output almost doubled in the Aramco-run Tanura refinery, but *decreased* from 2.6 million barrels to 1.1 million barrels between 1969 and 1974 at the Petromin-run Jeddah refinery, although expansion of this refinery had been planned. This was minimally compensated for by the opening of the 15,000 barrels per day refinery at Riyadh. Asphalt production, an important material in road building, more than

---

1. Saudi Arabia, *Development Plan 1975-1980*, P.114

doubled to reach 6,700 barrels a day. The Petrolube plant was opened in 1973 and produced 75,000 barrels per day. Fertilizer production increased and the notorious Jeddah steel rolling mill reached 22,000 tons a year production as double shift operations were started in 1975, still far from the original target, particularly as the planned second stage of the mill was not completed.

Most of the Petromin sponsored projects under the Plan involved surveys of potential mineral resources and petrochemicals industries and pipelines, estimated to cost SR4,549 million, while the allocation for construction of approved projects was only SR548.8 million which was meant to cover expansion of the Jeddah refinery, the second stage of the rolling mill and office buildings for Petromin, as well as land purchases for future plants and the activities of the Sulphur and Tanker companies. Petromin carried out a number of studies for the future use of gas and the creation of new hydrocarbon based industries, as well as surveys of mineral resources for future exploitation.

The non-oil based industries' rate of growth was in some sectors higher than planned: benefitting from the construction boom, cement production grew at 16.7 per cent annually, to reach 1,150,000 tons in 1974 and with a capacity of 1.5 million tons. Other products, however, did not reach their target of growth rate. Employment in manufacturing grew from 36,100 in 1970 to 46,500 in 1975, but most of this was in small workshops employing less than 10 people which cannot be described as factories.

The failure of other industries is reflected in the vagueness of the description of their achievements in the Second Development Plan which mentions that:

> 'Several new industries were founded and others expanded during the first plan period to the point that there are now substantial operations in: — basic construction material and products — furniture — glass containers — food processing — beverages — textiles and apparel — paper products — plastic products — industrial gases — paints — household detergents — printing and publishing.'[2]

In other words, the list of industrial projects was almost identical to that of the pre-Plan period, and there had been little new investment, leaving non-oil industry still on a largely artisanal scale, except for a few cases of foreign investment and repackaging of imports.

---

2. *ibid*. p. 192.

Table 3

Employment in Manufacturing Industry

Number of employees/company		Number of companies	
	1965	1967/8	1971
1 person		4,394	4,251
2 to 4 people	7,509	3,769	4,308
5 to 9 people		729	746
10 to 19 people		169	194
10 to 49 people	140	79	105
50 to 99 people		12	16
100 and over	30	17	18
Total	7,679	9,174	9,638

Source: Adapted from Ramon Knauerhause, *The Saudi Arabian Economy*, Praeger Books, 1975.

According to a different source[3], by 1974 non-oil industry included 9,360 establishments employing 36,012 people i.e. an average of 3.8 people per establishment. The assets of these companies amounted to SR716 million, i.e. an average of a mere SR7,600 per firm. Of these 3,563 were involved in textiles, clothing, leather and leather products and employed 5,959 people, i.e. 2.8 people per production unit; 2,526 firms were involved in food processing, beverages and tobacco, employing 10,601 people, i.e. 4.2 per factory; 1,474 firms were in wood and wood products such as furniture and employed 4,429 people, or 3 per factory. 864 companies were in metal products, machinery and equipment employing 4,260 people, 5 per factory; and 793 firms were in non-metallic mineral products employing 6,065 people, or 7.6 per factory.

The general picture is one of a highly undeveloped industry, extremely dependent on imports, and with little prospect of lessening that dependence, since hardly any local materials are available for use in production. The food industries rely on imported wheat and soft-drinks concentrate, the furniture industry on imported wood and sheet metal, and the textiles on imported textiles. The building industry is living through a boom period which may not last too long.

3. *Financial Times*, Saudi Arabia supplement, 12.1.76.

By the end of the Plan period 261 new licenses had been issued for new industries, which may be a positive indication for the future and studies had been completed for the construction of 3 grain silos, flour mills and feed-milling complexes, whose main use was mainly to be the storage of imported supplies.

Predictably commerce increased satisfactorily by an average of 12.6 per cent as a result of the expanded financial resources and the import of basic commodities and luxury goods. Food imports, for example, had risen from a value of SR1,011 million in 1970 to SR1,745 million in 1973 while perfumes and cosmetics imports rose from SR13 million in 1970 to SR41 million in 1973.[4]

The field where most development took place was infrastructure, primarily transportation, communications and construction. During the Plan period 3,221km of new roads were completed and another 900km rebuilt, representing 79 per cent of the target of 4,312km. The shortfall was due to the need to import materials, delays in beginning work and increased costs. The major ports were expanded, giving Jeddah 10 new berths and Dammam 7 new ones. Feasibility studies were carried out for the expansion of the smaller Red Sea ports such as Yenbo, and Jubail on the East coast which were all due to be implemented during the Second Plan. A programme of airport improvement was started during this time and will be implemented to improve the large airports and build 7 wholly new ones in towns which haven't yet got one. The use of the railroad increased as the expansion of the economy called for greater transport facilities, despite the difficulties inherent in the railway system.

The Plan's targets in telecommunications, with automatic telephones and satellite connections were not fulfilled and work was concentrated in the large population centres at the expense of the isolated rural areas. The original plans also proved to be too limited and therefore had to be revised. Under the Second Plan new projects are being prepared. Postal services improved from their previous chaotic level. Under the Plan the Electricity Services Department was created to improve electricity supplies which were both unreliable and totally heterogenous: residential and commercial consumers were offered different voltages and frequencies, and distribution has been extremely unreliable. The new

4. *Saudi Arabia Monetary Agency*, Annual Report 1976.

department was meant to standardise the service and extend it to reach more of the population, as only 2.2 million people had any electricity services whatever in 1974, which, once again, means that the rural population were deprived to the advantage of those living in larger cities.

The First Plan's achievements in the health sector were limited by difficulties of insufficient staffing, lack of organisation and coordination between the different departments, and the minute number of local staff, trained at home or abroad, as well as port congestion. Despite this the number of doctors and nurses increased considerably: altogether 15 hospitals were built and the number of hospital beds increased by about 500. Towards the end of the Plan period, certain measures were taken to improve general organisational standards and they may assist the implementation of the Second Plan.

In conclusion it seems that the First Development Plan was by no means an unqualified failure, as it was an improvement on the previously chaotic situation. It started the establishment of the planning processes, collecting reliable statistics and created the possibility of measuring the degree of progress, defined in terms of modernisation. However, given the regime's secrecy concerning real population figures, it is unlikely that any amount of statistical improvements could reach an adequate level of reliability, making it doubtful whether the existing administration would be able to implement the next plan, regardless of its sophistication.

At the level of infrastructure, the Plan did make some significant progress though doubts must be cast on the consultants responsible for irresponsible advice on such issues as the needed expansion of telecommunications and other projects. But in industry the Plan's failure was largely due to developing social and economic trends which are accelerated by the Plan's options, rather than countered by them. Therefore investment remained low except in small-scale projects, where financial results were rapid, such as bottled drinks and small assembly plants for the building trade boom. The local bourgeoisie's investment policy is determined by immediate profitability, which is greater in trade. This group is not concerned with primitive capital accumulation, a process which is considered by many economists to be an essential beginning to the development of a balanced economy. Immediate profitability suggests investment abroad, or in speculation and trade, none of which contribute to the productive capacity of the national economy. The government has tried to encourage industrial investment on the part of private business, but the incentives it offers have been insufficient to turn substantial amounts of capital away from trade. The trend towards commercial capitalism is a result of the particular circumstances of Saudi Arabia's economic transformation: the country is not transforming its mode of production through its own process of industrialisation and the internal development of its own natural resources. On the contrary its natural resources, primarily oil, are being extracted and exported in an unprocessed condition, and the 'development' of the country is taking place as a side-effect of the income derived by the Saudi Arabian state from the export of oil. Therefore the main source of capital derives from the state's conscious policy of setting up a private sector and its diversion of substantial amounts of its income to private business. This takes place mainly in the import business and in the payment of 'commissions' to middlemen, and in the long run may be aimed at creating a class of wealthy business people dependent on the regime for their continued prosperity.

The agricultural policy which emerges from the Plan appears to be based on the importation of techniques and principles which may have been successful in the USA where conditions are totally different, but seem to have little relevance to Saudi Arabia. Highly industrialised methods have been introduced in agriculture, encouraging the depopulation of the countryside, and the deterioration of the soil and the possible exhaustion of the already limited water supplies. Mechanisation and capitalisation have encouraged the concentration of agricultural units and the exclusion of the small peasant. The new agriculture which may be developing is of great benefit to importers of tractors, water pumps, fertilizers, new seeds and machinery, but it is unlikely to benefit the Saudi Arabian agricultural population (see below Chapter VIII).

The difficulties and problems which have arisen in the conception and implementation of the First Development Plan are also likely to arise in the Second for a number of reasons: both were devised with the assistance of US consultants, both are a collection of piecemeal suggestions, both fail to take into consideration the restrictions imposed by an inadequate infrastructure, both fail to recognise the limitations imposed by the inexperience of the administration and the lack of manpower at all levels.

THE SECOND DEVELOPMENT PLAN 1975-80

A mere two months after Faisal's assassination, Saudi Arabia once again hit the headlines in the West when details of the Second Development Plan for the period 1975-80 were announced. Unlike the first occasion, when reactions were a product of fear and concern for the 'stability' of the most important oil-exporting state, the announcement of the $142 billion investment plan for the next five years created enthusiasm among Western exporters who proceeded to treat this plan as an invitation to a bonanza of superprofits.

This ambitious plan reflects many of the contradictions inherent in Saudi Arabia's society and economy. Its introduction states as 'fundamental values and principles which guide Saudi Arabia's balanced development' objectives which include: 'to maintain the religious and moral values of Islam'.[5] The Plan does not, however, go on to explain how this is to be achieved and the specific programmes discussed in the following 600-odd pages read like a blue print for the creation of a dependent capitalist economy with no reference to the relationship between the society implied by such a model of development and the values of Islam, in the broadest sense, let alone in its puritanical Wahhabi variety.

The next goal is to 'assure the defence and internal security of the Kingdom'. The government's expenditure on arms in recent years has made it clear that this is a major priority in Saudi Arabian policies, but its place in the Development Plan as a goal seems odd, where it gets no further discussion and when, in any case, the contribution of defence to development is highly questionable.

The next two goals are to 'maintain a high rate of economic growth by developing economic resources' and to 'reduce economic dependence on export of crude oil'. These goals raise the whole question of the development options selected by the regime. Their implementation implies considerable diversification out of oil exports and into the local processing and use of oil in hydrocarbon industries and along with this the slowing down of

---

5. Saudi Arabia, *Development Plan 1975-1980* pages 1 for this and following quotes.

extraction to maximise income and lengthen the time span during which the country is able to benefit from its oil. As we shall see the Plan makes provision for the initiation of hydrocarbon based industry, which will take many years to come into operation and, as it is being designed and built by western contractors, there can be no guarantee that Saudi Arabia is being sold the technology most appropriate for its specific conditions. As we have seen,[6] foreign companies seem to be more concerned with a quick sale than with the quality and long term durability of their products. As for the conservation of resources for the future, the policies of the regime both before the announcement of this Plan and since, seem clearly to contradict this objective. The regime's pathetic concern for the welfare of the advanced capitalist world has led it to fight a rearguard action in OPEC to keep oil prices down (thus bringing about increased production and sales to achieve a similar level of income) while at the same time extracting and exporting oil in quantities far greater than those needed to generate the income necessary for the implementation of the Plan as well as amply providing for the extravagances considered essential to the lifestyle of the House of Saud.

The final three goals of the Plan express the continuing intention to create a basic material and human infrastructure considered appropriate for the 20th. Century, in a country which only 20 years ago was totally undeveloped. The Plan aims to 'develop human resources by education, training and raising standards of health' — certainly necessary objectives in a country whose population is still largely illiterate, lacking skills appropriate to modern technology, and whose health standards are far behind the capacity of the state's income to meet. The Plan aims to 'increase the well-being of all groups within the society and foster social stability under the circumstances of rapid social change'. Increased well-being is to be achieved through the provision of health and education facilities, and of welfare payments distributed by means test to those who can prove that they have no other adequate source of income. The problems brought about by rapid social change are not discussed in the Plan although they are of prime significance to the people of the country and their future. Indeed it seems that few of the policies devised and

---

6. See chapter III, page 55 on SAFCO and chapter IV, page 86 on the television system.

implemented give any consideration to the social upheavals they are causing or to the nature of the society which these projects are likely to bring about. The aim to 'develop the physical infrastructure to support achievement of the above goals' includes the building of roads and the improvement of all communications. Despite considerable building programmes in recent years there is not yet a basic, let alone an adequate, infrastructure for a modern economy.

The ordering of priorities is revealing in that defence and the preservation of moral values are given top priority but remain ignored in the rest of the Plan while the projects aimed at increasing the well being of all groups and the development of a human infrastructure all involve aspects which are liable to lead to social instability.

At the purely practical level, listing the development of physical infrastructure as the last and least significant item seems unrealistic given the importance of this sector. It involves building everything from roads to hospitals as 80 per cent of the capital investment in all the sectors of the Plan will go into construction since at this stage most of the health, education, social and industrial projects are essentially building plans. Since local industry was unable to supply more than a fragment of the needs of the building trade before the Plan, this means that practically every material needed will have to be imported. In 1976 it became quite clear that the port and road facilities were unable to cope with the level of imports despite the expansion which took place during the First Plan. Port surcharges in Jeddah reached 85 per cent and 45 per cent in Dammam and waiting periods for ships to unload could be up to six months! As a result the government hired Carson's, a private helicopter company, to unload cement at Jeddah port at enormous cost. The weakness of Saudi Arabia's infrastructure is the first reason why the Plan can hardly be achieved within the given time span.

A second reason is the extent of the Plan's ambitions, which involves an investment of 10 times the amount of the First Plan, which would require a highly efficient administration capable of multiplying its activities by that factor or the importation of people with such skills. The lack of qualified and unqualified workers in all fields also means that it is unlikely that the Plan can be implemented even if large numbers of building companies import everyone from their unskilled labourers to their top designers. The level of investment plans also leaves the Saudi Arabians responsible for running the Plan open to pressures from expatriate firms trying to sell a product, as even when these administrators are highly qualified they will not have the time and opportunity to study proposals in sufficient detail.

Finally the way the Plan was devised, with the advice and assistance of three different US concerns, the Stanford Research Institute, Harvard University and the Arthur D. Little consultancy organisation, gives some explanation of the orientation of the Plan. Geared substantially towards the importation of sophisticated highly technologically advanced processes which are not necessarily appropriate to the local situation, the main priority of the Plan appears to be the 'recycling' of as many petrodollars as possible. Some departments were just told to send in plans for anything they can think of, and the Plan itself was not devised as a single coherent unit which would ensure coordination and cooperation between sectors whenever possible.

The discrepancy of the Plan's goals and the reality of its content was clearly pointed out by a Saudi Arabian businessman in a conversation with a British journalist:

> 'If the first two priorities on the list are obeyed, then they defeat all the others. How can a nation, determined to preserve Islam and its own internal security at the same time, throw open its doors to foreign technology, manpower and all other kinds of influences? Who is going to run all these industries and services when they are finished? Look at it this way, there are a maximum of 4 million Saudis in this country. Discount half of them because they are women, another 25 per cent are children, which leaves a potential workforce of 1.5 million. After manning the Armed Forces, the National Guard and the police, how many Saudi are there left to run the government and private sectors? Saudi Arabia has opted for conventional industrialisation on a large scale and meantime the Western countries are doing a good job making sure the money they have spent on oil gets back to them.'[7]

7. *Middle East Economic Digest,* Saudi Arabia, special report, December 1976 p.32.

Table 4. THE PLAN'S ESTIMATED RECURRENT AND PROJECT COSTS [SR millions].

	Recurrent costs[1]	% total	Project costs[2]	% total	Projects as % of total projects	Total
Economic resource development	4,518.5	4.9	87,616.5	95.1	26.4	92,135.0
Human resource development	43,907.3	54.2	36,216.5	45.2	10.9	80,123.9
Social development	18,148.8	54.6	15,064.0	45.4	4.5	33,212.8
Physical infrastructure development	12,530.8	11.1	100,413.8	88.9	30.3	112,944.6
Subtotal, development	79,105.4	24.8	239,310.9	75.2	(72.2)	318,416.3
Administration	18,010.6	47.2	20,168.6	52.8	6.1	38,179.2
Defense	14,652.8	18.7	63,503.7	81.2	19.3	78,156.5
External assistance, emergency funds, food subsidies and general reserve	54,857.9	86.4	8,620.3	13.6	2.6	63,478.2
Subtotal, other	87,521.3	48.7	92,292.6	51.3	(27.8)	179,813.9
Total	**166,626.7**	**33.4**	**331,603.5**	**66.6**	**100.0**	**498,230.2**

[1] covers items under Chapters I, II, III, of annual budget (i.e. salaries and allowances, supplies and services, transfers subsidies, foreign aid). [2] covers items under Chapter IV of annual budget together with public corporation project costs and public financing of private sector and joint venture investments normally carried in Chapter III of annual budget.

Overall view of the Plan

Before discussing the different sectors of the Plan it is worthwhile taking an overall view of its divisions and allocations. As can be seen from Table 4 on page 152, project expenditure represents 66.6 per cent of expenditure and recurrent costs only 33.4 per cent but the distribution of expenditure within the projects budget gives a different picture, with the lion's share going to physical infrastructure (30.3 per cent) followed by economic resources (26.4 per cent) which includes all economic projects including oil-related ones, and defence taking the third place with 19.2 per cent although presumably some of the physical infrastructure projects are also of military value.

If one looks further at the main development programmes as they are outlined in the Plan, it can be seen that certain sectors, such as agriculture which could lead to economic self-sufficiency receive a very low share of the allocations, although they could be expected to take priority in a country trying to develop itself and progress towards an independent economy.

Table 5, which presents the main development programmes in greater detail than Table 4 can be expected to give a clearer picture of the development policies being implemented. Here, the major item of expenditure is education, taking 23 per cent of development allocations, followed by expenditure in local municipalities (16.7 per cent) which can include all local authority services from sewage to offices and houses, and then industry (14 per cent) and water desalination (10.7 per cent). If, on the other hand, we look at other projects and their costs by comparison with the above, it is to be noticed that defence is allocated a sum representing 24.5 per cent of 'development' allocation, and the undefined heading of 'funds' receives 20 per cent without explanation in the text.

The investment programme is ten times larger than that in the First Development Plan at 1974-75 prices, on which the entire Plan is devised. It allows for 15 per cent inflation per year, which is quite unrealistic considering the real rate of inflation which has been at least 40 per cent excluding housing, for which costs during 1975 and 1976 rose at a geometric rather than arithmetic rate. In 1977 it seems that the rate of inflation is being brought down slightly and that the housing situation has improved as a result of the rent regulations introduced in 1976.

Table 5

Main Development Programmes

Project	SR millions	%
Water and desalination	34,065	10.7
Agriculture	4,685	1.5
Electricity	6,240	2.0
Manufacturing and minerals	45,058	14.1
Education	74,161	23.3
Health	17,302	5.4
Social programmes and youth welfare	14,649	4.6
Roads, ports and railroads	21,283	6.7
Civil aviation and SAUDIA	14,845	4.7
Telecommunications and posts	4,225	1.3
Municipalities	53,328	16.7
Housing	14,263	4.5
Holy Cities and the Hajj	5,000	1.6
subtotal	**309,104**	
Other development	9,312	2.9
Subtotal development	**318,416**	**100.0**
Defence	78,157	
General administra-	38,179	
Funds	63,478	
Subtotal other	**179,814**	
Total Plan	**498,230**	

Source: *Saudi Arabia Five Year Plan 1975-1980,* p. 602.

The Plan's ambition can to some extent be measured by the fact that it calls for an annual growth rate of 10.2 per cent of GDP in real terms overall, including a projected 9.7 per cent annual increase in oil-related industries, which is probably an underestimate, and 15 per cent each for health and education, while a mere 4 per cent growth is expected of agriculture. The volume of construction is planned to increase by 60 per cent per annum over the entire period!

Although the Plan formally calls for diversification from a single crop economy to one of greater self-sufficiency, the relative position of the oil sector is only expected to decline from 86.6 per cent in 1975 to 82.1 per cent of GDP in 1980 and, although the non-oil private sector's growth rate is supposed to increase from 11 per cent to 14.9 per cent over the whole period, some sectors, such as agriculture, vital to any prospect of economic independence, are given low priority. On the whole, the Plan appears to be a first serious attempt at creating a basis for a modern economy in Saudi Arabia, although the degree of independence and the viability of such an economy are doubtful.

Divided into four major sections, economic resources, human resources, social development and physical development, the Plan gives details of all major projects of Saudi Arabia's planned social and economic changes in the coming five years or in some cases for a longer period. Following closely the organisation of the Plan document, we will look at projects for water supply, agriculture, oil and non-oil industry, education, health, manpower and physical infrastructure.

Water

As the Plan points out, in a country as short of water as Saudi Arabia, the supply of water is the first problem to be solved in practically any economic or social venture. In the past, in the scramble to get water by powerful people and institutions, irreplaceable sources were wasted, and as a result local agriculture was ruined by the introduction of mechanical water pumps and the lowering of the water table. This is particularly clear in the case of the Wadi Fatima region in the Hejaz, where since the 1940s a water company has made contracts with both local spring owners and built pipelines to supply Jeddah with water. The local nomads and peasants were unaware of the quantities of water which would be piped and of the likelihood of depletion, as the consumption rate involved in the cities was beyond the scope of their imagination.

'The water crisis stemmed from the fact that during the past twelve years many wells and three major pipelines were constructed in the *wadi* without any survey being made to determine specifically the location and quantity of water. When they made agreements with the water company, most of the Bedouins who owned springs felt certain that there really was enough water underground to meet the needs of both the farmers and the water company. Not only the *wadi*, but Jidda and Makka as well, will face serious difficulties if the water reserves become completely depleted in the years ahead.'[8]

The Plan does not, however, discuss these problems, but merely lists various means by which increased water supplies will be found. Although some very large underground water reserves have been discovered in Najd, and there are massive desalination projects, the natural resources are not being given the care they merit.

The Plan predicts an increase from the 1974 daily supply of 6,681 thousand cubic meters to a 1980 output of 10,100 thousand cubic meters per day. This is to be achieved by continued search for underground water resources. High water consumption industries are to be restricted to the coasts where desalination plants can be installed, and agricultural expansion limited to areas where proven long-term water resources exist. It involves the drilling of new wells and transmission systems, and large numbers of desalination plants producing both water and electricity.

One of the problems of water management is that the control of the water supplies is administered by a number of different authorities including the Ministry of Agriculture and Water, the Municipalities Department of the Ministry of the Interior, the Water Desalination Organisation, whose different plans are to be coordinated by yet another Ministry, the Ministry of Planning. Furthermore, despite talk of conservation and management of resources giving due consideration to the level of reserves, in practice wells continue to be drilled regardless of any investigation of consequences and water depletion is a serious danger.

Agriculture

Although the Plan recognises some of the problems inherent in its policies, the projects which have been decided for the Five Year period take little consideration of these difficulties. The

---

8. Motoko Katakura, *Bedouin Village* University of Tokyo Press, 1977 p.33.

difficulties which the Plan recognises are the following:
'—Over-exploitation of the range caused by increasing use of motor vehicles.

— High rates of water extraction by pumping, resulting in falling water tables and increasing salinity, and the abandonment and relocation of farms.

— Mining of non-recharging aquifers to meet the demands created by urbanisation and industrial and agricultural expansion, before the structure and capacity of these aquifers have been adequately defined and a long-range National Water Plan has been developed.

— Reduced consumption of some traditional crops — dates in particular, but also sorghum and millet — as prosperity leads to changes in consumer diets.

— Slow emergence of commercial agriculture because of the relative attractiveness of other sectors for entrepreneurial talent.'[9]

The objectives of the agricultural policy are 'to raise the per capita income and improve the welfare of rural people, minimise the Kingdom's dependence on imported food and release surplus labour for employment in other sectors.'[10] The regime plans to increase production by mechanising agriculture and encouraging private enterprise on a large scale 'water resources permitting' (!), while at the same time it is reducing the rural population to meet the demand for unskilled labour power in the cities. Apart from statements on planned production targets for 1980, there are among the projects listed a series of feasibility studies and the creation of national parks (!) which will be built, designed and organised under the auspices of the Saudi-Arabian-US Joint Commission on Economic Cooperation. Are we to see a copy of the Yellowstone Park in Saudi Arabia? The Plan also intends to improve training in agricultural methods of a greater number of people, including a two-year technical course to produce 385 graduates by 1980 and summer training for 1,500 university students, whose enthusiasm for agriculture is unknown.

The Saudi Arabian Agricultural Bank, which was set up in April 1963 with an initial capital of SR30 million, is expected to expand and to provide larger amounts of credit for the new class of

---

9. Saudi Arabia, *Development Plan 1975-1980*, p.123-4.
10. *ibid.* p.124.

capitalist farmers to mechanise, buy new varieties of seeds and improve their marketing and processing facilities. The total sum allocated for this purpose during the Plan period is SR427.7 million.

The policies implied by the Plan are an intensification of the existing trends of concentrating production in fewer hands and the elimination of the traditional small peasantry. A fuller discussion of agricultural policies and of nomadism can be found in the following chapter.

Oil and Minerals

Apart from the finalisation of the government's takeover of Aramco, plans for oil development continue to involve further exploration, research and training. Production levels are not considered to be the responsibility of the Plan.

The emphasis put on the search for other minerals means that the regime's understanding of diversification does not involve the domestic development of local resources, apart from the search for other minerals which could be exported to the advanced capitalist states in an unprocessed condition. In the First Plan, and even more in the Second, a significant proportion of allocations has gone to mineral research and exploration. The allocation for the Plan period is SR777.4 million to the Directorate General of Mineral Resources. Research and exploration are to continue to be carried out in association with the US Geological Survey and the French Bureau de Recherches Géologiques and Minières.

In mining ventures, Petromin has reserved the right to take a 50 per cent stake in any commercial venture which may arise from exploration programmes, and a clause to this effect has been entered into each contract signed. By 1977 as well as the two above-mentioned exploration companies, the British Steel corporation (Overseas), and Rio Tinto Zinc of the UK were carrying out general surveys while a number of other companies had already obtained exploration licences. These include Granges of Sweden, exploring for phosphates in the Turaif region, the British firm Consolidated Goldfields searching for gold in the Mahd ad Dhabab area in the northern Hejaz; as noted above a gold mine was operated there between the early 1930s and 1954 when reserves were considered to be too small for continued exploitation. The Canadian firm Noranda Exploration is looking for copper, zinc and nickel at Kutum-Talah in Asir.

As in other sectors, the government's policies are favourable to foreign investment, despite the 50 per cent Petromin stake which has to be negotiated in detail when an exploration licence is under discussion. Foreign partners are given a 5 year tax holiday and are allowed to repatriate capital and profits without restriction. In mining, the regime's logic for joint ventures is somewhat different from that in industry. In this case the government thinks that the foreign partners have to be given a stake in order to give them the incentive to participate in these mineral ventures, although the regime might well find that foreign companies could still be persuaded to participate on a contractual basis. Certainly the oil companies did not abandon their projects when they lost formal ownership of the concessions.

Heavy industry

The Plan's heavy industry projects are based on the creation of a vast gas gathering scheme which aims to gather 5,000 million cubic feet a day, most of which up to now has been flared, i.e. wasted. The project is being executed by Aramco and was designed by the US Fluor Corporation. In mid-1977 it was estimated to cost $14 billion. The gas will be gathered from a number of oil fields, separated into different qualities and piped to electrical power stations, export terminals and the two industrial complexes which are to become major petrochemical centres, Jubail on the Gulf Coast and Yenbo on the Red Sea coast. To bring the gas to Yenbo will involve construction of an 823 miles pipeline, longer than the Alaskan pipeline.

The Jubail complex on the Gulf is to be the larger of the two industrial centres. Original projections for it were grandiose and, since the Plan was announced in mid-1975, the various projects which compose it have been rescheduled. The first stage installation of gas gathering was, in mid-1977, due to be completed by 1981, while the rest of the scheme was to be staggered well into the 1980s. The project involves the building of a port of at least 29 berths, a city with all the necessary facilities for 175,000 people, and heavy industry including four petrochemical complexes, two export refineries, four fertilizer plants, an aluminium and a steel plant to produce 3.5 million tons per year. The entire project is to be managed by Bechtel Corporation, but as well as rescheduling, other problems have arisen.

For example, Mitsubishi Heavy Industries of Japan, which was committed to building a 300,000 tons a year ethylene plant tried to cancel their plans in 1976 as they feared losing money: from early 1975 to late 1976 the cost of their unit had escalated from $346 million to $1,700 million. However, in mid-1977 as a result of pressure from the Saudi Arabian government and fear of loss of status for other projects of theirs in the country, Mitsubishi made an arrangement with the Japanese Government agency, the Overseas Cooperation Fund, by which the financial loss would be shared equally between the two organisations and the project was resumed.

The second major industrial complex is to be located at Yenbo, 200 miles north of Jeddah on the Red Sea coast, a small traditional port. This port will be developed thanks to oil and gas pipelines coming from the Gulf region, which will supply an export refinery and a petrochemical complex. On the assumption that the Suez Canal remains open, this would cut the journey to Europe by 5,000km both for oil and gas, and could be a commercial proposition, despite the fact that large tankers cannot go through the Canal.

Although these status projects will be considerably delayed and at best only implemented in the mid-1980s, they are not likely to be cancelled altogether as the present regime considers them of importance and a matter of prestige, and would lose face if it cancelled them. The projects do raise a number of problems. There is first the possibility that the local production of petrochemicals may exceed Gulf and World needs by the end of the 1980s if no further downstream industries are established in Saudi Arabia to absorb the production. This problem is increased by the fact that Saudi Arabia is not the only state in the region with such projects: Iran and the United Arab Emirates are planning similar industrial complexes, and although Iran's domestic consumption should easily use its production the same is true neither in Saudi Arabia nor in the Lower Gulf States. The 'danger' of Saudi Arabia flooding the world with cut price petrochemicals seems unlikely, as they are likely to be high cost producers and on past form the plants are unlikely to operate at full capacity.

Heavy industry projects in Saudi Arabia in the past have suffered from design and construction problems: both the Jeddah steel rolling mill (see p.138) and the SAFCO plant in al Hasa (see p.56) have been failures, the first apparently because of the absence of experienced management and skilled labour and the second thanks to incompetent design. The type of problems inherent in these two cases are certain to reproduce themselves, and there is little the government can do about unscrupulous contractors who concentrate on cheap building techniques. The authorities have taken some action by encouraging the development of joint ventures. Unlike other third world countries where this is done because the local government is financially unable to set up such operations on its own, in Saudi Arabia the aim is different. Joint ventures have the advantage that should something go wrong in building, design or operation, the foreign company which is involved has to assume part of the responsibility. This is an incentive to encourage such companies to ensure that design and plans are suitable to local conditions and the factory will be able to operate according to plans once completed, otherwise the company will suffer financial loss and have to make the necessary improvements and repairs. It is hoped that this policy will ensure that the industrial complexes in Jubail and Yenbo turn out better than the Jeddah steel mill, and also that the petrochemical plants are built with greater care for safety precautions, given the inherent dangers of such plants.

Non-oil based industry

In this field the government hopes that private enterprise will dominate. As noted above, Saudi businessmen have not been very keen on productive industry projects for a number of reasons: first they can make far higher profits much faster by investing in trade, specifically in the import business, or in real estate and housing, where costs and profits have been soaring; second, foreign competition has meant that most local products are unlikely to be competitive, especially since there is no tradition of industry. In fields such as clothing, the cultural trend is in favour of western styles, rather than 'ethnic' products, except for men's wear.

Despite this context, the Plan aims at developing industry through the private sector and concentration is on building materials, agricultural inputs, food processing and basic household goods, in other words 'import substitutes' or activities linked to building. Private business can obtain loans for investment from the Saudi Industrial Development Fund, which started operations in October 1974. These loans can reach up to 50 per cent of the capital requirement at 2 per cent interest. Between its creation and January 1977 the SIDF received 287 applications for loans, 192 of which were projects for building materials.

The other main government encouragement to industries comes in the form of industrial zones. Provided with electricity, water, sewerage and roads, these estates are rented out to companies at nominal rents, and electricity and water provided at subsidised rates. All these facilities are difficult to get at all anywhere else and this is likely to result in a concentration of industry in these zones, located at Jeddah, Riyadh, Dammam, Buraidah, and Mecca.

By mid-1976 the government had received so many applications for building materials projects that it issued 100 licences and the Ministry of Industry stated that no further licences for cement block manufacture would be issued as capacity would meet demand till 1980. The Plan calls for production of cement to rise from 1,450,000 tons per year to 10,360,000 tons per year. In 1977 plants for galvanised iron, aluminium sections, bricks, reinforced tiles, prefabricated building materials and houses were being built. Alireza-Sumitomo were building a steel pipeline mill to supply the Aramco market.

Outside building materials various larger scale industrial projects have been mooted since the announcement of the Plan. Ghaith Pharaon's REDEC company is planning a joint venture fertilizer plant in the East using the facilities created by the general expansion of the region, and a number of motor assembly plants have been suggested. General Motors wanted to assemble 7,000 cars and light trucks a year, a Saudi-Japanese project was on the drawing board, and Juffali and Daimler Benz had got together to assemble 15-ton trucks. After the Saudi Industrial Development Fund had turned down all these projects on the grounds that their costs would be too high and that their products would not be competitive with imports, most of them were dropped. In 1977 the Juffali-Daimler Benz plant was under construction despite this setback, and the partnership re-applied for a loan, but even if that were refused, the company could rely on the German partners to assist the viability of the project.

In food-related industries, the Plan calls for the building of grain silos to a capacity of 210,000 metric tons by 1980, of flour mills to produce 1,350 tons a day. Meat processing and dairy product plants are also underway, and the food-related sector has attracted some interest among local investors, although some of the dairy projects are run under contract by British and other foreign companies.

In the non-oil sector, there are few cases of investment outside the food processing and building-related industries. Without clear encouragement by the regime, including positive tariffs against competitive imports, and the lack of industrial projects based on local resources and a minimum of imports, the present trend of final assembly and minimal industrialisation will continue.

Education

The Plan aims to increase schooling to cover enrollment of all boys aged 7 and 8 and 90 per cent of those aged 6, and to provide school places for 50 per cent of girls between 6 and 12 by the end of the Plan. At the intermediate level the Plan aims for enrollment of 95 per cent of all male and 80 per cent of female elementary school graduates by 1980 while at secondary school level, 50 per cent of boy and girl intermediary graduates are to be enrolled.

The development of teacher training is to raise the number of Saudi Arabian qualified teachers graduating from teacher training colleges from 4,547 in 1975 to 7,914 in 1980 for the lower levels of education, and from 102 in 1975 to 3,233 in 1980 for further education. The number of university graduates is to rise from 1,703 in 1975 to 7,158 in 1980.

It seems clear that what is going well in the educational field is the building of schoolrooms, which is supposed to average 90 a week completed during the 5 year period. Many of these rooms however, when built, will remain empty as there are no teachers to work in them or houses for the teachers to live in. The number of indigenous teachers is still very small by comparison with expatriates (see p.81). A syllabus has yet to be devised which can raise future educational standards and provide the formation of a population educated in the skills necessary for the development of the country. Although it has been modified in recent years, it still relies largely on the Egyptian model combined with religious training, with no regard for the type of society and economy which other sectors of the Development Plan seem to be leading to.

Health

Like education, the health services are mainly staffed by expatriates, and here Egyptians and Pakistanis dominate the field.

Table 6. Estimated Manpower by Occupational Group 1395-1400 (Thousands)

Occupational group	1395 Saudi	1395 non-Saudi	1400 Saudi	1400 non-Saudi	Increase Saudi	Increase non-Saudi
Managers, officials	7.4	6.3	8.7	12.4	1.3	6.1
Professionals	48.4	15.7	52.9	23.5	4.5	7.8
Technicians and sub-professionals	25.0	31.4	33.4	81.3	8.4	49.9
Clerical workers	67.5	31.4	99.6	121.8	32.1	90.4
Sales workers	82.3	47.1	97.2	112.6	14.9	65.5
Service workers	105.2	47.1	134.5	145.2	29.3	98.1
Operatives	40.0	25.1	57.1	51.4	17.1	26.3
Skilled workers	70.1	47.1	93.5	101.9	23.4	54.8
Semi-skilled workers	170.0	62.8	265.0	162.5	95.0	99.7
Unskilled workers	244.0		296.4		52.4	
Farmers	311.2		281.0		-30.2	
Bedouins	114.9		98.7		-16.2	
TOTAL	1,286.0	314.0	1,518.0	812.6	232.0	498.6

Five Year Development Plan p. 247.8

The Plan aims to ensure that by 1980 there will be one doctor for every 2,000 people in the country, but it is unlikely that many of these doctors will be Saudi Arabian, local training facilities are still at an elementary stage and only 1,061 scholarships will be available for "medical and related personnel". Once again the Plan is concentrating on building and aims to provide 2.5 hospital beds per 1,000 population, and to increase and upgrade a number of dispensaries and health centres. The Ministry of Health is being allocated SR 12,338.1 million for project expenditure and SR 4,963.5 million for recurrent expenditure.

The labour force

Given the magnitude of the Plan and the general expansion of all levels of administration and other activities, the expected increase in the demand for labour is enormous. According to the Plan the Saudi Arabian labour force in 1975 was 1,286,000 including 27,000 women. The figure of 314,000 presented for foreign labour is totally unrealistic and presumably excludes the Yemenis, officially said to number 250,000 but in reality they may number up to 2 million; it is also unclear how many Yemenis may be included in the figures for Saudi Arabians, as until 1972 they entered the country without any records being kept of their movements.

The expected rise in the labour force is illustrated in Table 6 but this is one occasion on which official figures are suspect given the regime's secretive attitude to the number of foreigners resident in the country. The table does however give an indication of the distribution of labour within the economy: apart from the traditional sectors of farming and herding (probably overestimated to give the impression of a higher national population) the categories given are difficult to divide between the primary, secondary and tertiary sectors, but it does appear that there may be a preponderance of service workers and administrators over productive workers, i.e. that the tertiary sector has a disproportionate role in the economy. This is not surprising since the country's bureaucracy is relatively large, and the main economic sector is oil, which employs few people, followed by trade which employs many.

The Plan estimates that the labour force will increase by 730,600. Migrant workers will form 498,000 of these, and according to these figures, by 1980 foreigners will account for 35 per cent of the total even according to official figures. Let us here repeat that in this case official figures are almost certainly an understatement of reality. The increase of the Saudi Arabian labour force is to be 3.1 per cent per year, amounting to 232,000 while the proportion of Saudi Arabian women in the workforce is to rise by 3.3 per cent yearly.

Although it is clear that a large part of the labour force necessary today is important to the creation of the physical and social infrastructure, as well as the various showpiece industrial projects, most of the foreigners involved will be repatriated to their countries of origin when work is completed. At that time a labour force will be necessary to run the plants, operate hospitals and teach in schools; even the most capital intensive petrochemical plants will need workers. This means there will be an increased need for a local labour force of all degrees of skills. If, however, the regime sets up industries which are highly capital intensive, as would be reasonable in a country which has a population as small as Saudi Arabia, the need will be mainly for highly skilled personnel.

It is to fill the gap of a skilled labour force that the Plan calls for the expansion of higher and technical education as well as of vocational courses for Saudi Arabians. As we have seen above there is little enthusiasm among young Saudi Arabians for vocational training, even among the poorer sectors of society and among those who have left the rural areas to become urban dwellers and workers as these people consider manual labour to be degrading. In the past, this has meant that the Vocational Training Centres have been unable to fill their quotas for students. The Plan, while discussing objectives for increased vocational training and the creation of new courses, says nothing about how students are to be attracted to the centres. In 1975 the existing centres had a capacity of 2,500 trainees, and the Plan aims to increase the capacity to 5,200. But by 1975 the total number of people who had graduated from these centres was only 4,371. This leads one to believe that the supply of skilled and semi-skilled labour is not likely to improve significantly in coming years, and that migrant workers will continue to find work in Saudi Arabia.

In administration, the implementation of the Plan calls for an increase of 170,636 employees over 5 years, or an annual increase of 18 per cent among civil servants. This will take a considerable number of the graduates from secondary schools and universities, thus excluding from the productive sector many people who could contribute to its establishment on a sounder basis. Most qualified people who do not take up positions in the Civil Service are likely to opt for the trade and import sector where possibilities for enrichment will appear more attractive, as commissions are high for agents and middlemen, and trade is regarded as an honourable calling. Except for a few exceptional cases of far sighted people, few willingly enter industry or skilled trades.

Physical infrastructure.

Despite the efforts of the First Development Plan, the condition of the physical infrastructure still leaves much to be desired. In discussing the ovjectives for the Second Plan, it is worth looking at the level of achievement of the First Plan.

The country's size must not be underestimated when discussing infrastructure: from Dammam on the Gulf to Jeddah, as the crow flies, is over 1000 km and from Turaif near the Jordanian border to the Rub' Al Khali is over 1600 km. The First Plan aimed to build 5,212 km of roads for which SR 2,538 million was allocated, i.e. the cost per kilometer was estimated at SR 487,000. In fact, only 4,129 km were built at a total cost of SR 4,029 million, i.e. a cost of SR 976,000 per kilometer, almost double. The Second Plan selects the following factors as partial causes for this increase in cost:

> '— Delays occurred between budgeting and actual start of work.
> — Many types of essential services, materials, and equipment increased in cost.
> — A general lack of construction manpower, materials, equipment and spare parts — or delays in acquiring the manpower and supplies.'[11]

Certainly, whatever bottlenecks and other cost-increasing factors affected the First Plan, will only get worse during the Second. Among other factors, western companies are increasing their prices according to the inflation rates of their own countries, as well as that in Saudi Arabia.

The Second Plan calls for the building of 13,066 km of paved

---

11 *ibid.* p.495.

roads and 10,250 of earth surfaced roads. The allocated cost per km is SR 750,000 which, judging from previous experience is likely to be exceeded, particularly as costs have rocketed since early 1975.

Concerning ports, the expansion of Jeddah and Dammam is to continue during the Second Plan. At Jeddah 18 new berths are to be built and 16 at Dammam. Three other ports are to be greatly expanded; Yenbo will get another 14 berths for the new industrial complex and Jubail will get a completely new port while Jizan will get 2 new berths. The planned expansion of ports, although it appears to be extremely necessary in the late 1970s, while a massive construction boom is in progress, may in twenty years time turn out to be well beyond the lasting needs of the country for its trade once the economy has settled down.

The existing major airports are also being expanded to take wide bodied aircraft, and new airports designed or initiated. Although it is not mentioned in the Plan, in 1977 the idea of reopening the Hejaz railway was once again raised. Since this question has come under discussion at regular intervals since the end of the first World War and nothing has ever been done, it is unlikely that the situation will change in the 1980s.

Telecommunications are due for considerable expansion under the Plan. From 94,000 telephone lines in 1974, capacity is to be expanded to 666,000 lines of which 490,000 are to be in service by 1980, providing 20 telephones per 100 residents in the larger cities and 5 per 100 in the small centres. Postal services are to be improved and postal staff trained, while communications satellites are also to be established.

The Municipalities Department which is responsible for local government, has been allocated 4.5 per cent of the development budget for the improvement of basic facilities such as water supplies, sewers and the construction of municipal offices, the paving of roads, street lighting etc... By the end of the Plan, 162 full municipalities should have been set up, i.e. villages or settlements upgraded to the title of municipality thanks to the installation of basic service facilities and an element of town planning.

The housing sector reveals some of the major deficiencies of the present situation. As the Plan points out, in the First Plan period about 75,000 standard dwellings were built while the need was more than double that, i.e. 154,000. The result was overcrowding and shanty towns. The Second Plan aims to increase the housing programme so that 225,600 houses are to be built, 122,100 of them by the private sector. This is clearly a long-term objective as in the 1975-1980 period only 52,500 public sector houses are to be built.

Since the private sector is only too willing to make vast profits on private housing with the aid of loans from the Saudi Real Estate Development Fund, public sector housing is mainly restricted to the lower income population. But despite the very modest aim of 52,500 houses in the five year period of the Plan, it was only well into 1976 that the first contracts were awarded for the building of 41,750 houses in a number of towns: Riyadh, Jeddah, Dammam, al Khobar, Mecca, Medina and al Khafji.

The private sector, on the other hand, is flourishing. Rents in recent years have rocketed and the government has been compelled to control them and, in 1976, to freeze rents. In 1976 villas suitable for expatriate executives were being let at $30,000 a year! Thanks to these exorbitant rents private builders have not hesitated to obtain loans from the Saudi Real Estate Development Fund, which by 1977 had already financed the building of 100,000 units compared with the 40,000 envisaged by the Plan.

These absurd rents mainly affected expatriates, but they also had a serious impact on the indigenous population, as the trend towards urbanisation and the destruction of old houses to make way for roads and other facilities, drove a lot of people on to the housing market. Many of them could not conceivably pay the kind of rents being asked for in the private sector, while public sector housing was totally inadequate. The situation eased off somewhat in 1977 as a result of two factors: first the government has insisted that in all contracts involving investment of over SR100,000 the contractor must provide housing for all imported workers, from managers on down, as well as their own office space. The second factor has been the 1976 rent freeze which combined with expansion of housing to ease the situation.

Conclusion

As we have mentioned earlier there are two great gaps in the Plan's details: defence and oil. Defence was expected to absorb 20 per cent of the budget but clearly it absorbed more as can be seen from the massive defence expenditure of 1975 and 1976, worth SR 23,723.7 million in 1975-76, i.e. 21.4 per cent of the budget and SR31,906.4 million in 1976-77, i.e. 28.8 per cent of the budget.

By contrast in other sectors actual expenditure has usually been below allocations, because of the administration's inability to fulfill its commitments. The gap between appropriations and expenditure is gradually closing, but, as was pointed out by Planning Minister Hisham Nazer in his dicussion of 1976-77 expenditure "I don't know whether it is because we have been more efficient or because of the rate of inflation"[12].

Escalation of costs between 1975 and 1977 was staggering. By February 1977 the Saudi Arabian government felt obliged to take action against companies who presented inflated tenders, and cancelled a number of bids when it was discovered that for four electrification projects estimated by the initial consultants to cost £700 million inflated tenders had been sent in by 8 western firms; the lowest bid was twice as much as the estimates which had been revised to take account of inflation. The companies involved were given a warning and excluded from the second round of bidding which did not in fact take place as the government decided to share out the contracts among four friendly third world countries, India, Pakistan, Taiwan and South Korea.

Only two months later, another highly inflated bid was revealed, this time for telephone equipment: Philips had put in a bid which was a cool $6 billion over the Saudi Arabian projected cost of $1 billion for the same project. Exclusive negotiations were broken off and the bid was rejected personally by Crown Prince Fahd. This caused some public interest as Fahd's own son, Mohammed, was due to get a commission totalling more than $100 million out of the deal. This case combined two hot issues in Saudi Arabia in 1977: inflated tenders from western companies, and the problem of pay-offs to middlemen. According to one business person close to these negotiations: "It must have been a painful moment for Fahd, but he wanted to set the example that would show that the time of wild spending had to come to an end."[13] He presumably also wanted to make a point about commissions to intermediaries, since the regime has stated its intention of reforming the regulations which compel any foreign company operating in Saudi Arabia to have a local agent.

This law was originally meant to ensure Saudi Arabian

---

12. *Financial Times,* 8.7.1977.
13. *Washington Post,* 2.5.1977.

participation in the development of the country, rather than leaving it all to expatriate firms. But the importance of trade in the economy has meant that many people restrict their activities to receiving agent's fees and commissions on imported goods without actively participating in the business.

In 1977 it is difficult to estimate the extent to which the Plan will achieve its stated objectives, or the effect this will have on the future of Saudi Arabia. But certain aspects are becoming clear: Saudi Arabia has become less willing to allow western consultancy and contracting companies to make massive profits at the expense of the country's development. The regime is therefore increasing its economic links with other foreign firms as it is unable to fulfil its Plan on its own, and to do this it has turned towards some of the more advanced third world countries.

The Plan's objectives concerning human development can only constitute an improvement on the previous situation, as the development of education and health facilities are positive factors.

Should the industrialisation policy succeed and oil be kept in the ground, the country could establish a viable economy well into the 21st Century, on the basis of hydro-carbon industries advantageously placed through exhaustion in other countries. The country could then supply many parts of the world with plastics, fibres etc. This will only happen, however, if oil resources are conserved and the policy implemented with care. As for other industries, unless serious steps are taken to develop an industrial sector using raw materials and manned by personnel able to deal with the problems of advanced technology, the country will remain dependent on advanced capitalist states.

CHAPTER VIII
SAUDI ARABIA
SOCIETY IN TRANSITION

The political and economic upheavals we have discussed above could not possibly have taken place without a parallel transformation of the country's social structure. Although in the 1920s the regime initiated attempts to reshape society, it is only in the 1960s and 1970s that traditional social structures have been uprooted under the contradictory pressures of a traditionalist political system and economic policies based on western concepts of development and modernisation. Since the mid-1960s the destruction of traditional relations has been accelerated to the extent to which, by the late 1970s, it could be said that only insignificant traces of the traditional structure remain. But it is only in the 1970s that new structures are beginning to emerge, slowly and uncertain of their direction.

One of the contradictory forces affecting this process is the ideological distortion of traditional nomadic tribal ideology for political ends. The regime's existing political structure, based on family rule with a few concessions to western educated technocrats, could not be maintained without this transformed tribal ideology which, with the addition of religion, gives the regime a legitimacy based on the rewriting of history.

The other prevalent force is western influence. This is most significant among those men who have returned with high qualifications from the west, and favour a transformation of the country modelled on the USA. These 'modernisers' support a capital intensive industrial programme and a way of life based on the nuclear family villa with all modern conveniences.

In this chapter we will look at the process of social transformation, at what has happened to the traditional groups, the *bedu* and the agricultural communities, and at the development of new social formations, i.e. at the birth of a class structure.

The Nomads

When discussing the changes which have taken place in the social conditions of the nomads, it is important to remember their special position in the ideology of Saudi Arabia. On the one hand *bedu* tribal values are hailed throughout the state as an ideal to which all should aspire, while on the other, the state was established on the basis of distortion and manipulation of these social and political values.

The traditional nomadic political economy was based on tribal divisions, armed conflict and nomadic pressures on settled agricultural communities. To establish overall control of the Arabian interior, Abdel Aziz undermined this system by forbidding inter-tribal struggle and making the tribal leaders dependent on him for subsidies. At first presented as a form of tribute to the leaders, these subsidies soon became their only source of income, and therefore their only means of control over their tribes. They therefore soon became dependent on the House of Saud who controlled the income, which was external to the traditional structure: it first came from British subsidies and later increased considerably once oil was discovered. Because Abdel Aziz had obtained international recognition of his supremacy in the region, all income was paid to the Crown, and tribal leaders had no alternative but to take the subsidies which allowed them a minimum of local power within their communities.

The ideological relationship between *bedu* values and the regime is reflected nationally at the social, economic and political levels. A number of the contemporary contradictory features of Saudi Arabian society arise from the mis-adaptation of these values to a structure which has been superimposed on them and is fundamentally different. For example, the general assumption that all Saudi Arabians have the right to an official position and a regular government income regardless of that person's individual contribution to the country's development is an extention of the system of subsidies. The converse of this is the value ascribed to self-sufficiency and independence of the individual, which is reflected in the fact that welfare benefits are selective and means tested. In 1972, 4 per cent of the population received social security benefits but the system of distribution was corrupted by 'nepotism' in many cases. Again, what we in the west consider to be nepotism is understood in the Saudi Arabian context as the traditional fundamental duty of any individual to look after the members of his family and tribe before anyone else.

At the political level, confusion arises from the large number of official and unofficial networks which are used to contact the higher levels of power: each tribal leader, or even village chief, tends to assert his right of access to the King on even the most minor issue. As a result, the established government structures and hierarchies are ignored, both on national and local issues, and administrative confusion reigns. The King's *majlis*, by contrast, is no longer the open meeting to which every man, whatever his social status, can come and have access to the King: now it is little more than a royal court meeting where only the privileged have any real access, and the ordinary individual may wait for years to get an audience, or have to grease a few palms on the way.

These are merely a few examples of the way in which *bedu* ideology has been adapted and distorted as a result of the upheavals of this century. This distortion of values serves a specific ideological purpose, that of maintenance of the political system frozen at a point in the transition between tribal rule and modern political structures. This ideological distortion of *bedu* values could not have happened if the mode of production had remained intact. But one of Abdel Aziz's first decisions was to abolish intertribal raids: taking place after trade had declined, this decision reduced the nomads to herding as their only economic resource, and initiated the destruction of the traditional method of distribution.

The last thirty years have seen a number of radical transformations in the conditions under which nomadic herding have taken place. The introduction of motorised transport has made is possible for herders to take their livestock further away from sources of water, as water could be brought to them by tanker lorry, and the digging of wells along oil pipelines and elsewhere has reduced the need for long-distance travel between winter and summer grazing lands, thus providing the opportunity for sheep and cattle herding and the creation of mini-settlements with basic facilities.

At first sight it seems that the increase in the number of wells, the development of motorisation and the change from camel herding to cattle and sheep herding are positive contributions to the Saudi Arabian economy by making it possible to maintain a traditional economic activity, transform it and adapt it to the new circumstances which have arisen from the oil-export economy. But such a view ignores a number of serious problems: motorisation and the increase in the number of wells have made it possible for larger herds to be maintained on the same area. The larger herds are using up grazing land much faster, and their rapid transportation from one grazing land to another leads to overgrazing.

It is currently estimated that if the process of overgrazing is not to lead to total desertification, already a serious danger, grazing pressure shoud be reduced by about 50%, which means a reduction of livestock of the same proportions. The alternative is the deterioration of this fragile steppe land into irretrievable desert.

Between 1957 and 1965 another serious drought contributed to the increased indebtedness of the nomads towards local usurers, and combined with the nomads' pre-existing poverty (SR 350 per capita income in 1967) by comparison with urban dwellers, to encourage the *bedu* to abandon the nomadic way of life. At the same time the creation of a consumer market has attracted many young nomads to leave their families and seek work in the towns, thus further affecting the viability of the nomadic communities.

These factors are bringing about an increasingly rapid rate of sedentarisation, estimated at 2 per cent of the nomad population per year in the mid-1970s. For example, in 1969, in the Garith, Sulays Bishak and Khamis Mushayt regions of Asir 13,500 men between the ages of 15 and 55, out of a population of 37,500 men of this age group, had abandoned nomadism and left the region. The problem of overgrazing accentuates the present socio-economic trend of sedentarisation of the nomadic population which still forms about 10 per cent of the national population, while 20 years ago it formed about 50 per cent of it. The rapid disappearance of the nomads leaves the herding areas to the small group of entrepreneurs who are setting up 'ranches' for the large scale production of meat for the urban markets.

For while small scale herding is in decline a form of ranching is being developed by those who have invested in means of transport and are increasing their herds. This option is open to those who, owning enough livestock to begin with, have been able to invest with loans obtained from the Agricultural Bank, by calling up their credit with poorer members of the community and by buying trucks to carry water to livestock. These entrepreneurs are running profitable businesses and are expanding, thus further endangering the range through overgrazing.

Among the small scale herding families, a process of disintegration is taking place: as the family gets poorer the incentive for migration becomes greater and soon the younger men will leave for the towns where they hope to earn enough to pay the family's debts and to acquire some of the consumer goods they desire. While this is happening the herding unit is left in the care of the older men and women who, on their own, are unable to sustain an economically viable production unit as the range is barren due to overgrazing and their poverty forbids them access to motorised transportation. They therefore become more indebted or have to settle somewhere where they have access to the state welfare structure.

The young destitute nomad has a number of options: he can continue in herding by becoming a 'cowboy' for the entrepreneur who is expanding locally, or he can migrate to the cities where he becomes a wage labourer, seeking to acquire consumer goods such as transistor radios and cars. In the cities the young *bedu* feels all the poorer since wealth has become a sign of status: poverty which used to be respected, particularly among nomads who were all more or less poor, has now become a stigma in a society where the display of wealth is conspicuous and where considerable pressure is applied by importing companies to expand the consumer market.

Young nomads from certain highly respected tribes join the National Guard where they will be well paid, will learn skills and have a place of honour within Saudi Arabian society commensurate with the traditions of nomadism. Their income will enable them to assist their dependent families.

While young men go and seek their fortunes in the cities or in the National Guard, the rest of the family finds itself in increasing economic and ideological difficulties. For old people the economic difficulties can only be alleviated by welfare payments as they are unable to adapt to the urban pace of life, while at the ideological level, all members of the family have become aware of the fact that traditional herding is fast disappearing and that the present generation of nomads is probably the last one. The younger men who have migrated have had to take unskilled jobs or work as taxi drivers as they are mostly illiterate. The *bedu* family thus realises that the only future for the younger generation lies in education which will enable the children to gain access to some of the wealth which is available in the kingdom. Education will also help the young ensure a living for their parents in old age.

This search for education is driving many nomadic families to centres which have sprung up more or less spontaneously. These centres have schools and sometimes clinics, a mosque and a shop. These embryonic villages are becoming the instruments for the transition process between nomadism and sedentary life. As well as providing health and educational facilities, the villages also help nomads make the socio-psychological transition between nomadic herding and the tribal solidarity associated with it, and the urban context, where relations are impersonal and exclude these links. In the village nomads can live a semi-nomadic existence, the administration is run by members of the same tribe who are sympathetic to the problems of its members and the children can be taught the skills needed to cope with this new world without being separated from their families, while the parents are also introduced to the new way of life.

As well as the unplanned changes which are destroying the traditional nomadic political economy, there have been a number of formal official attempts to settle the nomads. The precursor of these schemes was the creation of the Ikhwan agricultural settlements under Abdel Aziz, but most of these fell apart when the movement was crushed in 1930.

King Faisal once again initiated a settlement policy in the 1960s. The official explanation for this move was very similar to that used in other countries, where regimes, be they right or left wing, are bent on settling their nomads. The only exception is Mongolia where the nomadic mode of production has been retained while services such as education and health have been organised to fit in with nomadic patterns. Although it could be said that conditions in that country are better because of the quality of the range and the availability of water, the policy of settling nomads always means the waste of an economic resource, as the range used by nomads cannot be effectively used in any other economic pursuit. As we have seen, ranching wears out the range and leads to desertification. In Saudi

Arabia in particular, it would be easy for the regime to invest in mobile health and education facilities for the nomads. But apart from Mongolia, all regimes, whatever their political colour, seem to consider the settlement of nomads to represent a form of progress.

Reasons for putting pressure on nomads to settle are numerous: at the political level it is easier for a regime to exercise sovereignty over settled than over nomadic peoples. Most regimes consider that the settlement of nomads will bring about greater national cohesion in new states. Nomads are supposed to benefit because it will raise their standard of living by providing secure employment and access to modern facilities and tools for their work. They will participate in the economic development of the Kingdom by helping to increase agricultural production both for food and as raw material for industry. In fact in Saudi Arabia, most nomads who settle do not go into agriculture but into the cities where they become labourers and the benefits they obtain from settlement are purely financial. Socially the change in their situation leads to psychological confusion as to their real role in society.

The government projects for the settlement of nomads have concentrated on turning them into peasants, a scheme highly unlikely to succeed given the mutual hostility that has traditionally existed between nomad and peasant. The failure of these schemes has not resulted in the maintenance of herding but in the more of less direct migration from herding to cities. As in so many other cases in Saudi Arabia, rather than set up a number of projects with an integrated and careful approach, the government has resorted to large scale showpieces.

The first official settlement project was initiated in 1961 in the Wadi Sirhan in the north west, near the Jordanian border. There were good strategic reasons for establishing a settled community there as it is a border area in a region where Saudi control had in the past been contested. The drought which began in the late 1950s was used as a pretext for the project. At first the government provided drought relief to the nomads, and in 1961 it set up the agricultural project. To encourage the *bedu* to settle, the government provided them with water pumps, tractors, manual tools and seeds, as well as fuel, all of which were meant to be used collectively. The lack of detailed plans and studies of the likely effect of such a project or of the way it should be implemented to fit in with the local ecological and social environment resulted in its failure. As the drought receded in the 1960s, many of the former nomads returned to herding as their experience of agriculture had been negative: they were far away from a marketing centre, were settled on land which had little water and high salinity, and consequently produced little. The sharing of tools etc. went counter to the individualism of nomads and as the possibilities of herding once again seemed promising, more and more of them returned to the nomadic way of life. The project gradually declined and in 1972 it was abandoned, the equipment being put up for auction as a final admission of failure.

The showpiece *bedu* settlement project is known as the King Faisal Model Settlement Project at Haradh. Located about 250km south east of Riyadh, this project was initiated in 1964 and aimed at using some of the newly discovered sources of underground water to turn into agricultural land a region which had not previously been cultivated. The stated objectives of the project were to take a step towards Saudi Arabia's agricultural self-sufficiency as well as to improve the social and economic status of the al-Murrah tribesmen who live in the region, and to train them in skills which will be useful in the new economy.

Within the project area 1,000 families were to be settled in agriculture and supplied with health and educational facilities. The 45,000 dunum included a training centre and experimental farm, cultural and commercial centres and 40,000 dunum divided into 1,000 farm plots of 40 dunum each (200 x 200m), large enough to allow for mechanisation and for irrigation by the channels coming from 50 wells. The population was expected to rise to 8,000 people distributed between one large central village and 8 smaller ones, each including three types of residential accommodation as well as a mosque, a school, a park, shops and offices. Completed on schedule in 1971, the Haradh settlement has been a failure.

Initiated with the advice and assistance of the US Food Machinery Corporation, the project was set up on the basis of mechanisation and large scale planning which took no consideration of the social and economic traditions of the people who were being transplanted there, nor were the people ever consulted about their views on sedentary life. In the later stages of construction, some 200 al Murrah worked on the training farm. Although they were referred to as 'future, potential settlers' their treatment ensured that no al Murrah would ever want to settle there:

> '[the] al Murrah laborers lived six and eight together in bare concrete-floored and walled rooms approximately twenty feet by ten feet, with only the barest minimum of 'conveniences', none of them modern. Arranged barracks-style in long rows, these members of famous al Murrah lineages were joined by the vast majority of laborers at the project, most of whom came from the lower class sedentary folk from al Hasa, Oman, the Yemen or the Hadhramaut. Some of the al Murrah occasionally escaped these living quarters to stay in the tents of their relatives as they passed near Haradh or stopped for a while in the area during the course of their yearly migrations.'[1]

It is easy to imagine what they told their tribesmen about the settlement! The absolute ignorance and disrespect for tribal values expressed in the contractors' behaviour reveals the failure to plan according to the needs of the people. The planners involved obviously imagined they were building green suburbia with different houses for 3 classes of people, and a capitalist hierarchisation of the population, with a park and swings for the children to play in, while the mothers shopped in the shopping centre and the fathers worked in the fields! This in the middle of the Saudi Arabian steppe, with a people who have been renowned for centuries as trackers and camel herders! The absurdity needs no comment.

By the time the project was completed in 1971, it was impossible to persuade many al Murrah to settle there:

> 'Most [members of the actual sheikhly families of the al Murrah] were contemptuous of the way in which it was being run by the foreign contractors and of the low status that the al Murrah tribesmen had to take in this project, which had been announced to improve the nomads' way of life.'[2]

As not enough *bedu* were found to settle in the project, and those who did move in had not been properly trained to use the

1. D.P. Cole *Nomads of the Nomads,* Chicago, 1975, p.150.
2. *op.cit* p.151.

mechanical aids provided, the problems could not be ignored.

True to form, the government followed its usual pattern of action: it called in more foreign 'experts' who were equally unable to grasp the real problems of the situation, and attempted to hire a private agricultural company to set up a 15 year training programme for the nomads, with the intention of distributing the land at the end of that period. But as the returns on investment were predicted to be low, no agribusiness was prepared to take on the contract, and that included the companies which had been involved in the initial establishment of the project and had then acted with the tact and sensitivity for local values which we have seen. The government eventually decided to retain the experimental farm and turn the rest of the area into permanent grazing land to fatten sheep and cattle, activities which do not need a large number of settled people.

Official plans for the settlement of nomads in Saudi Arabia have failed, to some extent because of the insensitivity with which the projects have been carried out. But some informal settlements have succeeded and transformed some nomads into peasants and the process of transition is still continuing.[3] Although there seems to be little future for traditional nomadism in a nation that is being built on oil revenues, nomadism could continue to make a valuable contribution to the country's economy were the regime prepared to make some minor concessions in establishing mobile facilities. But as the situation stands, the regime is quite willing to continue to exploit *bedu* values at the ideological level to help in its control over the entire society, but at the same time continues to distrust the remaining nomads and wants to integrate them into the settled social, economic, and political fabric of the state. The current belief that nomads are excluded from society indicates the extent of social transformation in recent decades: not so long ago nomadic structures and the nomads were at the centre of the nation's life. Former nomads play an important role in contemporary society as members of the National Guard, as ranchers and in the various fields in which they have settled, while the remaining nomads eke out an ever more meagre existence in the desert and supplement

---

3. For a discussion of a slow and gradual transition from nomadism to agriculture in Wadi Fatima, read, M. Katakura, *Bedouin Village*, University of Tokyo Press, 1977.

their income with welfare payments. Overgrazing, which is unlikely to be prevented, presents the serious danger of desertification and the destruction of most of the Saudi Arabian land with economic potential.

Agriculture

In the 1960s agriculture was still estimated to occupy about 40 per cent of the total population, farming 0.2 per cent of the total land. Of the 525,000 hectares farmed, 20 per cent is rainfed, in Asir, the southern Hejaz and the Tihama coastal plain. The rest depends on oases, wells and springs in northern Hejaz near Mecca and Medina, Khaybar and Wadi Fatima, around the main cities in Najd, Riyadh, Ha'il and Buraida, the al Kharj agricultural project, and in al Hasa.

Land tenure traditionally was based primarily on small peasants owning their plots of land, but there were also larger landowners, either absentee landlords or local dignitaries who hired sharecroppers to look after their holdings: in the West and East about 50 per cent of holdings were rented, but in the north most of the land was owned by those who cultivated it.

The average size of holdings in the country is under 8 hectares, only one eighth of which is irrigated, but variations in size are considerable. In the south-west there are two main types: small holdings averaging 1.3ha, and larger holdings belonging to merchants and officials, whose average size is 4.5ha but can be as large as 100ha. These are cultivated either by sharecroppers or by agricultural workers under the supervision of a manager, when the farms are mechanised. In the eastern region, where many former peasants are working in the oil fields, some very small holdings are cultivated by sharecroppers: in Qatif, for example, the majority of holdings are under 1ha. In the north the average holding is 3ha but this includes both grazing and cultivated land.

In the 1960s the main crop was still dates, the only product which was ever produced in quantities large enough to allow exports. Grown mainly in the oases of the east and Najd, about 200,000 tons of dates were produced each year. Date production is however in decline as the general standard of living in the country improves and more food products are grown and imported at subsidised prices. Dates no longer form the staple of the diet. As the market has declined those owners of date palms who have turned to other economic activities have allowed the trees to be neglected.

The main grains grown are wheat, millet, sorghum and barley. Since the 1950s rice growing has been developed in al Hasa thanks to the development of an irrigation system. In Asir most crops can be grown thanks to the better climate, but production concentrates on fruit trees and vegetables as well as coffee and qat which cannot be grown anywhere else in the country. The following table gives some indication of agricultural production, the figures are for 1971-72 as no other year's data is available:

Agricultural production 1971-72

Product	Tons
wheat	38,954
sorghum	52,360
millet	17,244
barley	9,318
water melon	448,936
aubergine	8,256
tomatoes	110,950
dates	187,846
grapes	28,795
coffee	243
rice	68

Source: Kingdom of Saudi Arabia: *Central Department of Statistics*

As an indirect consequence of the exploitation of oil, the agricultural system has been changing and since the 1960s the process of transformation has accelerated. Holdings are being concentrated and the new larger holdings are being mechanised and used to develop new crops, while the small peasants are being squeezed out of agriculture altogether. There are a variety of pressures which explain the decline of the small peasant and the rise of the agricultural entrepreneur:

1. The introduction of mechanised model agricultural projects in traditional agricultural areas tends to cause a deterioration in ecological conditions which affects most seriously the small peasant: the use of mechanical water pumps rapidly leads to a lowering of the water table, this in turn means that the wells used by the traditional peasant become unusable. The lowering of the water table also results in a deterioration of the soil.

2. The low level of agricultural incomes (estimated in the 1960s to be about SR 300 a year) combines with the national labour shortage to encourage the migration of young peasants to the cities where they will find employment.

3. Traditional small holdings are poor and therefore present no attractions to the Agricultural Development Bank which is unwilling to make loans to peasants for improvements, crop changes or anything else as it considers small peasant units to be economically unviable.

4. The small peasant's inability to invest in mechanisation means that he becomes increasingly impoverished and indebted, and eventually may have to sell his land. This results in a concentration of land in the hands of small entrepreneurs who buy these small units, bring them together, mechanise and start producing for the market. They employ as wage labourers some of the peasants who they have helped to reduce to destitution. In recent years this process has been particularly acute in the Jizan region in Asir which was affected by drought and where the government has built a dam and irrigation system, thus assisting the development of capitalist agriculture.

5. The extreme poverty of the traditional agricultural sector means that agricultural income on a small holding is insufficient to maintain a family. As a result the head of a peasant family often tends to become a mini labour contractor: he decides which of his children will be sent out to work in the cities as labourers, which ones will be educated so they can get an administrative job and help the family later on, and which ones will stay in agriculture. This type of long term division of labour is necessary to ensure the basic survival income of the entire family, as agriculture on its own is unable to do so; it also sows the seed for the disintegration of the family unit.

6. The overall process of transition in Saudi Arabia is reflected in uncertainty about the future among the peasantry. On the one hand the number of peasants is declining, on the other agricultural methods are being improved for the larger units which remain. In al Hasa, for example, small holdings have been abandoned either because their owners have left to become labourers and hope to return when their economic situation is improved, or because these plots have been bought by workers who hope to settle there in their old age. The worker who comes from a peasant background retains his attachment to the soil, but the plots which are not farmed

deteriorate, thus affecting neighbouring plots, and leading to a general worsening of agricultural conditions.

As small scale agriculture declines and the impoverished peasant sells his plot, the tendency towards the concentration of agricultural land into fewer larger capitalist units intensifies. This concentration also contributes to the migration of the former peasants to the cities as larger agricultural units are more capital intensive: they cannot all become agricultural workers. A small entrepreneurial class is developing, composed of men who are able to obtain investment capital and to modernise agriculture, redirecting it towards the market, growing crops which can compete with imported foods.

The new agricultural entrepreneurial class is growing rapidly in the areas where the government has invested in infrastructure plans, such as the eastern region where the al Hasa irrigation and drainage project has reclaimed 200,000 dunums of land, improved drainage and irrigation, thus reducing the problem of ground salination. A similar situation exists around the newly-built Wadi Jizan dam in Asir and the irrigation networks set up in Qasim and Jawf.

The development of large mechanised capitalist agricultural units is encouraged by government policies. These policies derive from the national shortage of labour and the consequent desirability of encouraging former peasants to join the urban labour force, as well as to the government's stated ambition of achieving agricultural self-sufficiency and by general claims to 'development'. In the case of agriculture, as in other sectors of the economy, 'development' is interpreted as the introduction of agribusiness and the opportunity for some to make large profits, rather than in terms of developing the country's ability to be self-sufficient and rely on its own resources.[4] The government's model for agricultural development is based on the experience of a number of model farms which were set up in the country, and are mainly operated under the supervision and with the advice of western 'experts'.

The first agricultural model farm was established in the al Kharj oasis, 75km south-east of Riyadh. In the late 1930s, in order to satisfy the royal household's increased demand for better food,

---

4. For a discussion of the contribution of agribusiness to agricultural development in the Third world, read S. George *How the Other Half Dies*, Penguin Books, 1976.

the Minister of Finance, Abdullah Sulaiman al Hamdan, had a few water pumps installed and an irrigation canal cut to irrigate the land around the oases, and he hired an Iraqi engineer to operate the scheme. As a result of the developing relations with the US and the difficulties in supplies brought about by the war, a US agricultural mission visited the area in the early war years. Later, Aramco was asked by the King to provide water pumps, transportation and technology to modernise the farms. Their limited output led to dissatisfaction which brought about instant US Government response since the Americans feared disruption of their oil deals and wanted at that time to prove they were the good guys, and the British the bad guys in Saudi Arabia. In 1944 the US Government sent another mission which attempted to develop capital intensive agriculture in al Kharj. The missions were followed by US personnel hired by Aramco. On the al Kharj farms all possible varieties of grains and vegetables were grown at great cost, all for the use of the Royal Court in Riyadh.

Attempts were made to teach the new agricultural techniques to the local farmers who anyway resented the Americans' activities. This hostility, kept in check only by regular royal visits, was due both to traditional Wahhabi hostility to foreigners and to the American experts' patronising attitudes: they failed to recognise any value to traditional agricultural methods, and insisted that the only correct way to do anything was their way. It became rapidly clear that their methods were having a disastrous effect on the water table which was lowered as a result of the mechanical pumping, thus causing the ruin of many traditional farmers in the area who were soon reduced to becoming reluctant wage labourers on the American-run farms.

Encouraged by the so-called success of al Kharj, other model farms were started by Aramco, under the sponsorship of the King, in Hofuf, Qatif, Ta'if, and Wadi Fatima. These were all located in traditional agricultural regions and near large and expanding cities but, unlike al Kharj, the other model farms were not meant to maximise production on a large scale for the supply of neighbouring cities. They were meant to be experimental stations whose employees would focus their energies on improving local crops and agricultural methods by introducing mechanisation and encouraging the local agricultural community to adopt 'modern' agricultural methods. These projects were continued and expanded, and in the 1960s, two more experimental farms were opened, near Riyadh and Medina.

The government claims to want to convert the remaining peasants into modern farmers and sends 'extension officers' to the agricultural areas to demonstrate the qualities of 'improved' grains and other costly innovations, which are supposed to produce higher yields: the peasants' rightful scepticism at these innovations is interpreted by the salesmen as 'ignorance'. The result of this policy is that while some peasants become entrepreneurs and enlarge their holdings, use machinery and improved grains and fertilisers, traditional agriculture gradually disappears to be replaced by a few large scale projects. The traditional communities, and probably their socio-economic structures, are disappearing; even with the best will in the world and unlimited cash, it will not be possible to recreate them once the fundamental ecological structures of the oases have been undermined.

In areas where underground water resources have been discovered in reasonably adequate quantities, irrigation systems have been built and the land reclaimed for agriculture: this has been distributed in units of between 5 and 10ha to individuals and 400ha to companies who are willing to exploit the land with modern methods. Other irrigation projects, such as the Wadi Jizan dam, are aimed at improving water supplies in a region where water fall is comparatively high but erratic.

Another instrument for the modernisation of agriculture is the Agricultural Development Bank which provides interest free loans on short, medium and long-term agreements to farmers who are in a position to capitalise their operations. The loans are meant for the improvement of soil conditions, irrigation schemes, experimentation with new crops, new varieties of livestock, as well as the acquisition of machinery, equipment and fertilisers. By the very nature of its operations the Bank has no interest in lending to small peasants who are trying to eke out a bare living from the land. Run by bankers its terms of reference are profitability rather than personal satisfaction or mere survival, and its objective is to encourage the development of a capital intensive agriculture at the expense of the subsistence farmer.

The different facets of the agricultural development policies selected in Saudi Arabia combine to encourage the emigration of the smaller peasants from the rural regions and the concentration of agricultural land in the hands of larger landowners who have the capital to mechanise. The long-term effects of intensive agriculture on the soil and the water resources of the country are major questions which have not been given enough attention. It is however undeniable that the traditional agricultural structure is disappearing and the rural population declining. The existence of large aquifers may in future mean that new agricultural areas can be opened and the country increase its agricultural potential. It may even be able to feed its population, but this prospect is remote. It would certainly seem optimistic to suggest that Saudi Arabia should plan its post-oil economy on an agricultural basis.

The Formation of a Labour Force and the Migrant Population

The search for oil in the 1930s began a major transformation of the Arabian peninsula, but in the early years exploration in the Eastern region had little effect on the local population. A few men were hired as labourers with the surveying and drilling teams. But until the War for most of the labourers hired, employment with the oil company was simply an interlude in their lives of herding and agriculture. They were not integrated as wage workers dependent on their income from the oil company. Wage labour in either building or in the oil wells had little permanent attraction for either peasant or nomad, and it was only after the War, and particularly since the 1960s, that people were gradually driven to the oil wells as their major economic resource by the impossibility of living from their traditional economy in the new structure developing out of the state's oil income.

With the rapid expansion of the oil industry after the War, an increasing number of Saudi Arabians were hired by Aramco. After the initial years when the western capitalist working hours turned workers away from the industry, economic dependence forced them to adapt to the pre-planned working day. Long-term employment security acquired some significance for the workers and the creation of a stable labour force was also in the interests of the company which was, besides, under some pressure from the regime to provide jobs for the local population. The company set up training programmes from the late 1940s onwards, and their quality and standards have risen over the years.

As we have seen (p.95) there had been some labour unrest at Aramco in the 1950s, but since the company improved its working conditions and salaries Aramco workers have been comparatively

Length of service of Saudi Arabian employees in Aramco 1975

years of service	35+	30-34	25-29	20-24	15-19	10-14	5-9	0-4
No. of employees	94	953	3,015	2,427	716	428	285	6,620

Source: *Aramco Facts and Figures 1975*

privileged members of Saudi Arabian society, although there are few of them. One way in which the company saved itself money was by unloading some of their services and manufacturing requirements onto local entrepreneurs whom the company helped with capital and training. This meant that the entrepreneurs' workers were not liable for all the benefits of Aramco workers, thus decreasing Aramco's bills. Since the major strike in 1956 the only industrial action in Aramco was a canteen boycott in 1964 against the quality of the food and the standards of service.

Aramco was the first, and still is the only company which employs thousands of workers. The rest of industry is embryonic, except for building which has taken the bulk of workers outside the oil sector. On the whole, building projects did not at first create permanent employment and once the projects were completed, the indigenous workers tended to go home. Nowadays migrant workers move from one building site to the next.

Labour unrest outside Aramco, although it is badly documented, may reflect the relatively bad wages and conditions of workers outside the oil industry. In 1963 for example, workers at a cement factory went on strike to demand better medical care, housing, wages and training of Saudi Arabians to replace foreigners. The workers were forced back to work. In 1964 the police force struck in one province demanding pay increases and better housing, and seamen working at an oil terminal went on strike for overtime pay. Wages outside the oil industry have not kept up with the exorbitant rate of inflation. This, combined with the very heavy workloads imposed on the largely migrant labour force, has led to discontent.

Partly as a result of labour unrest, and partly under pressure from the increasing number of migrant labourers due to the building boom which began in the 1960s, the government issued new Labour Regulations in 1969, which were meant to replace the various decrees relating to labour relations which had been issued at various moments of crisis. Devised with the assistance of ILO officials, these regulations introduce provisions for groups of workers for whom there were previously none, such as juveniles, women and handicapped workers, although none of these groups forms a significant proportion of the labour force. They also create provisions for the training of workers and labour contracting. Rules are set out for hiring and firing workers, health and accident protection. In principle, these Regulations apply equally to Saudi Arabian and foreign workers, thus bringing them all under the same legislation and improving the conditions of immigrant workers. All private companies, including foreign ones, must treat their workers in the same way and provide the same facilities, and some degree of job security.

Of course these Regulations make no provisions for collective bargaining which is forbidden and punished by fines and imprisonment, as are strikes. If the culprits are migrant workers they are liable to be summarily deported. Should a worker have a grievance he is supposed to proceed by filing a complaint to his local labour office which will notify the worker's management. The case will then be heard at the labour office in the presence of a representative of the worker and of management. If no agreement can be reached the case is to be taken to a *Primary Commission* and in final appeal to a *Supreme Commission*. Claims must be individual, collective ones are forbidden.

By the mid-1970s, the labour force remained quite small. About 20,000 in oil and 40,000 in manufacturing, the rest of the embryonic 'working class' was primarily composed of migrant workers and involved in building. The building boom since the 1960s has been the biggest employer of local and immigrant labour. Former nomads and peasants forced into the towns to seek work may become the elements of a new working class, but this is unlikely as they form only a minority of workers in any industry and, although the present generation may be uneducated and work as unskilled labourers and porters in official establishments, their children are being educated and should qualify to join the lower echelons of the state administrative structure in ministries, schools and hospitals. However, education is not universal and there will be some uneducated people in the next generation.

The labour force, even when the migrants are included, is small and divided. The division is between the privileged workers in oil and the others. Non-oil workers have better conditions if they are Saudi Arabian than if they are migrant despite the Labour Regulations which, in any case, are not strictly applied. Migrant workers have the worst conditions, particularly at the lower levels, and those who work for particularly harsh employers, the South Korean in particular. Given that most of the migrant workers will not settle permanently in Saudi Arabia and are on short tours of duty, and that there are very few workers of Saudi Arabian origin, it is difficult to envisage the formation of a traditional working class in the country. If and when industrialisation gets off the ground, it seems likely that capital-intensive industries will be introduced. Therefore the few workers who will be involved in them are likely to be highly qualified and not to acquire the features normally associated with a working class. Those few people who remain uneducated and in search of unskilled jobs will be a minority.

Even before Saudi Arabia was unified as a state, the Saud regime was dependent on foreign workers. In Abdel Aziz's days, his advisers, doctors, and the teachers in the few existing schools came from Syria and Egypt. With economic expansion, and particularly in the 1970s with the current boom, Saudi Arabia's dependence on migrant labour has greatly increased.

The number of foreign workers in the country is a subject the regime is extremely touchy about. No official figures for foreign residents in the country are published: this is largely because the regime is concerned about the ratio of Saudi Arabians to foreigners in the country. Although the official population figures claim 7 million, it is generally believed that the native population is about 3,500,000. In 1975 official estimates of the number of foreign workers was 315,000. It is generally accepted that the number of foreign residents is at least 1 million.

In March 1977, the Financial Times listed the following: 500,000 Yemenis, 180,000 Egyptians, 88,000 Sudanese and Africans, 50,000 Pakistanis, 75,000 Indians, 30,000 Americans, 12,000 Britons, 30,000 other Europeans, 24,000 Koreans, 3,000 Indonesians, 'large numbers' of Palestinians and other Arabs.

These figures do not include Taiwanese or Japanese. The Financial Times' estimate of the number of Yemenis is questionable since, until recently, Yemenis have not had to comply with immigration procedures, and some of them have been settled in Saudi Arabia for decades; some even have their families with them. Other estimates of their number range from 1 to 2 million. In August 1977, President Ibrahim al Hamdi of the Yemen Arab Republic stated that there are 'two million Yemeni workers in Saudi Arabia, and they are of cardinal importance to that economy.'[5]

Foreign workers in Saudi Arabia can be found at all levels of the economy, ranging from unskilled building workers and porters to medical specialists and economic advisors. For example the late Anwar Ali, director of SAMA, was a Pakistani. There is an overall parallel between nationality and status in the workforce: the high level specialists tend to be from advanced capitalist states and the unskilled workers Arabs and Asians, but there are Arab technocrats and British lorry drivers.

Yemenis form the largest section. Their right to enter Saudi Arabia without immigrant procedures, which has been limited only recently, has contributed to the imprecision concerning their numbers in the country. Close to 2 million, mostly from the YAR, they are the only migrants who bring their families and settle in Saudi Arabia more or less permanently. But most of them come for a few years to earn money for their families who stay at home, and the workers return home for fairly long visits. Yemenis work mainly in unskilled jobs, primarily in building and portering. They are most numerous in the Western region, but there are significant numbers of them elsewhere.

Yemenis have even fewer rights than the indigenous population, as they have no access at all to the channels of power. Although officially they should receive the same benefits as Saudi Arabian workers, this is not often implemented and their access to health services is very limited. During the Yemeni Civil War between 1962 and 1967 they were often suspected of political activity, and large numbers of them were expelled without appeal. In 1967, after a period of political agitation in Saudi Arabia, 17 Yemeni workers were summarily and publicly executed although a Saudi Arabian organisation had claimed responsibility for the incidents of which they were accused.

In recent years, since the relationship between Saudi Arabia and

5. *The Middle East*, London August 1977.

the YAR has been good, and the need for workers in Saudi Arabia greater than ever, their numbers have been increasing. Saudi Arabian businessmen praise the qualities of Yemeni workers and some of them consider that Yemeni workers should receive all the social benefits that Saudi Arabians get, as well as training and educational opportunities. In this way they could become a skilled and stable labour force in the country and settle permanently.

There is also a small number of Yemenis of Hadhrami origin who have been living in Saudi Arabia for some decades. These people are mainly involved in trade and small businesses, and settled there after returning from South East Asia with some capital. These small traders are the people who will be most affected by the ban imposed in late 1976 on foreign nationals owning certain businesses and market stalls. Although this ban is selective and clearly does not apply to large scale joint ventures, it will hit hard the smaller Yemeni and Hadhrami shopkeepers.

Egyptians have been working in the Saudi Arabian educational system since the 1940s. Throughout the 1960s they played a dominant role in education, and in this way introduced some Arab nationalist ideas into the schools, causing concern to the regime, which tried to reduce its dependence on Egyptian teachers. This has not yet been achieved (see p.81). Egyptians also hold important positions in higher education and at various levels of the administration. The main priority for Egyptians, as for other workers in Saudi Arabia, is to make as much money as possible as fast as possible, to send it back home, and to survive in Saudi Arabia without suffering too much psychological damage. The remittances from workers in Saudi Arabia and the Gulf have become important sources of foreign exchange for the poorer Arab states.

Among Arab workers, Palestinians have a special place. Since 1948, large numbers of skilled Palestinians have come to live in Saudi Arabia where they have been working in the oil industry, the administration, education and social services, as well as in business. As an educated and politically conscious group, the Palestinians are particularly closely observed, and the authorities have no hesitation in expelling any Palestinian whose concern for his homeland goes beyond the limits acceptable to the regime. In 1955, about 100 Aramco workers who were Palestinians were expelled overnight because they were suspected of belonging to the Parti Populaire Syrien, a pan-Arab organisation. The Palestinians, as refugees with no homeland, are the only immigrants who expect to stay for an indefinite period and who bring their families with them if they have permanent employment.

Arab workers by and large should be more integrated into Saudi Arabian society by the mere fact that they share a common language, but except for a few, immigrants remain isolated from Saudi Arabian social life, and are by definition excluded from political activity. Other foreign workers who come on contracts, live and work in their isolated camps and never come into contact with the indigenous population, or even Arab migrants. This is true on the major South Korean sites and on some western sites: the migrant workers are kept away from the Saudi Arabian population and anyway have little interest in meeting them. Since there are no public cinemas or clubs, there are no local entertainments for the workers to go to.

Asian workers are increasingly numerous. First there are Pakistanis and Indians who are found in the medical services and also in unskilled labour, as well as in construction and engineering. A more recent trend is the influx of East Asians, whose numbers have taken off since 1975 and the start of the Second Development Plan. Since 1976 the regime, torn by the contradiction of needing workers while disliking the presence of foreigners, and particularly unbelievers, on its soil, has initiated a new system of contracts. Contractors who offer to bring all the labour they need with them, build and provide all the facilities necessary for these workers, and take the workers back after the completion of the contract, are given preference.

As a result the number of contracts allocated to Japanese, Taiwanese and South Korean firms has increased, and workers from these countries have arrived in large groups. For example in June 1976 the South Korean company Hyundai signed a $1 billion contract to build the industrial harbour in Jubail; for this project 5,000 Korean workers were imported. The same company also has a contract to build a naval base for the Saudi Forces. South Korean workers in Saudi Arabia work extremely long hours and their living conditions are a reproduction of conditions in Korea. Most of them are recently released soldiers:

'Once they get to the Middle East, these ex-soldiers will find life little different from that in the army. The workers are housed in isolated compounds built by the construction firms where they are supplied with Korean food, tapes of Korean TV shows and cassettes of Korean music. They are isolated from other workers, both from the indigenous population and from other countries, and from any danger of political "infection".[6]

Primarily concerned with saving as much money as possible Korean workers want to work overtime, to increase their wages, particularly as the wages paid by Hyundai are considered low, even by Korean standards. In March 1977 the Korean workers at Jubail rioted as a result of the low wages and difficult conditions. This caused serious concern to the Saudi Arabian administration which instructed the Korean government to improve conditions in its companies, or they would not get further contracts. Asian firms who enter into contracts which include labour have an advantage on European and American firms, as their labour costs are generally lower. For this reason they are so successful in penetrating the Saudi Arabian market for large construction contracts.

Foreign employees are also found in large numbers in the professional sector. Among them are industrial consultants from Western Europe and the USA and specialists ranging from engineers, teachers and medical specialists to geologists and television station personnel. As a rule these people come on short-term highly paid tax-free contracts and few of them are prepared to stay in Saudi Arabia for very long, as they find living conditions harsh both climatically and socially: the exorbitant price of black market alcohol and the absence of entertainments are facets of the 'quality of life' with which Westerners find difficulty.

Given the close relationship between the Saudi Arabian regime and the USA, Americans form the bulk of the highly paid professional group. Already 30,000, their number is expected to double by 1979, although recent American tax legislation may reduce this increase in favour of Europeans. Ranging from the US Army Corps of Engineers and other military advisors such as the Vinnell Corporation, to consultants on development policy working with the Ministry of Planning, medical experts working for the King Faisal

6. N. Disney, South Korean Workers in the Middle East, in MERIP no. 6, p.23.

Specialist Hospital, and many others, the American presence is undeniably the most conspicuous. Building contractors and businessmen seeking profitable deals complete the picture. There are English language schools in Jeddah and Riyadh catering for the children of westerners.

The biggest focus of American activity is of course the eastern region where Aramco has built its own town in Dhahran. This includes facilities such as golf course, putting green, baseball fields and tennis courts. The residential quarters are composed of American-style villas with watered front gardens, hedges and lawns. Aramco has its own television and radio stations which transmit in English and Arabic. The Aramco city is an extension of middle America in the Arabian desert.

As part of the European presence, Britain has the advantage of language. After Arabic, English is the second language in official and business circles, particularly since most of the educated cadres studied in the USA or Britain. British engineers, businessmen and other professionals play an important role, but the biggest contracts are for military assistance. Since 1975 there has been a new trend of importing British skilled labourers and dockers, some of whom find it difficult to withstand the conditions and quickly return: in the summer of 1976 the first batch of British dockers arrived under contract to help cut down the congestion of Jeddah port. Arriving in the middle of Ramadan, they found that they were expected to work a 12 hour day without eating and were unable to drink alcoholic drinks at night because they are banned. Within 10 days some of them were back in England, despite the fabulous wages they had been offered.

The other new trend in British-Saudi cooperation is the drain of British footballers and football managing staff who are leaving for Saudi Arabia and other Gulf States where they have been offered amazing salaries to coach local football teams. In the summer of 1976 the British football impresario Jimmy Hill signed a £25 million contract to build a new generation of Saudi Arabian footballers.

The Saudi Arabian regime's fear of foreign cultural influence may be exaggerated. Certainly the Europeans and Americans who come and work in Saudi Arabia have little influence on the national culture. Those who most come into contact with them are Saudi

Arabians who have been educated in the west and are already familiar with western culture. These people periodically visit the west and are already considerably westernised. The mass of the population is not affected.

On the other hand contact between manual labourers of different nationalities is limited by the language barrier, by their physical isolation from one another and by the fact that all of them want to earn much money and get out as fast as possible.

There is far greater danger of influence on the indigenous population by the Arab migrants who have closer contact with the people and who have comparable backgrounds and interests. This influence was seen at the political level in the 1960s when Egyptians and Yemenis were encouraging the indigenous population to adopt more nationalist political positions.

The government's present concern is with the number of migrant workers and their conditions. The last thing they want is a repetition of the unrest which took place in March 1977 on the Hyundai site, and conditions will be placed on contractors to avoid that level of dissatisfaction among the workers.

Official estimates of the demand for migrant labour in Saudi Arabia call for a 500,000 increase during the Second Development Plan period. This is clearly an underestimate. While the number of western workers (as opposed to professionals) may not rise, that of Asians is expected to continue its dramatic rise. Malaysian workers will join the Indonesians, Japanese and Koreans. Japan Airlines have plans to fly 300,000 workers to and from Saudi Arabia between 1976 and 1980, representing the arrival of 1,500 Asian workers a week in Saudi Arabia.

The Saudi Arabian Bureaucracy

Unlike other Middle Eastern countries, such as Egypt, Saudi Arabia does not have a long tradition of bureaucratic administration, as Ottoman penetration was never sufficiently far-reaching or long-lasting to establish one. But this is one field in which Saudi Arabia has caught up with, and overtaken, its competitors. Within just over a decade it has developed an almost totally immobile dinosaur, unable to reach and implement decisions on even the most minor issues.

The novelty of administrative positions gives their incumbents a secure and regular income which is not associated with any expectation of administrative efficiency. This is primarily due to the growth of administrative institutions alongside the perpetuation of the tribal-monarchic political structure: on the one hand the population uses the channels to which it is accustomed, i.e. the tribal leadership, rather than the new administrative offices, and on the other officials prefer to leave such a risky business as decision-making to the Minister, or preferably a member of the ruling family, however trivial or important the problem.

Having created the new bureaucratic apparatus, the regime itself is determined to retain the traditional relationship between the ruler and his subjects which induces local governors to try and overrule the official administration and appeal directly to the Court, so long as he happens to have good relations with the King or his entourage.

Many problems arise from overlapping between different levels and parallel responsibilities within the administration.

This is particularly true of local government, where there are already in principle two types of administration: on the one hand, the institutions of local government and on the other the local offices of the institutions of central government. The latter suffer from a number of problems, mainly the tendency to overcentralisation, the fact that lower grades do not have enough authority and need to refer every minor issue to the centre. All departments have a tendency to overstaffing. In his study of local government, al-Awaji eloquently introduces the problem of bureaucracy:

> 'Nothing is as unsystematic and confusing in Saudi Arabia as its local government, for the exact number and forms of major units of emarates are still unknown to many Saudi Arabians. Thus while some may still consider the geo-historical division (four provinces) as the basis of the present division, many people put the number at five. However, the official budget of 1970-71 classifies the major emarates into eight .... Provinces are divided into different smaller units such as districts or cities, in some cases, and only cities and villages in others. All divisions — regardless of size — are called emarates which are headed by umera. Although officially the amir of district, city or village is responsible to the above unit in his province, conformity to the rules is determined by the personality elements involved at any level. Direct contacts by an amir with the Minister of the

Interior or even with the King is a common practice.'[7]

This observation, although made in 1970, clearly shows that the Provincial Regulation Statute, announced in 1963 along with other reforms, had not been implemented. This is largely because the implementation of an efficient administration would mean the decay of the traditional relationship between the regime and the emirs of the different regions: since the stability of the Saudi regime is to some extent dependent on these people's support, it is hardly surprising that modern administration is not being encouraged. It is also relevant at this stage to point to the two factions within the ruling family, one of which tends towards support of modern structures and the other relying on support from tribal leaders and the traditional sections of the community. In these circumstances the duality of administration is merely one aspect of the on-going conflict between modernism and tradition.

Another major problem for the administration is the tendency to instantaneous and miraculous appearance of new institutions! This seems to happen whenever an institution is publicly exposed as scandalously inefficient: its chief administrators get together and decide that some changes are necessary. To plan them they promptly call in a firm of consultants, usually from the USA but sometimes from Europe, who arrive with their preconceived solutions and their total ignorance of and lack of interest in the dynamics of Saudi Arabian society. They then make recommendations which might be suitable in their country of origin but are in no way adapted to the culture of Saudi Arabia. Their flying visit usually brings about the creation of a new institution — and once they have left, things tend to carry on much as before. An example of this was the creation of the Supreme Planning Board in 1961, which was set up without warning or discussion about its role and power. After years of chaos, the Ford Foundation was called upon the study the problem and recommended some changes. As a result the Central Planning Organisation was founded to replace the earlier institution, but was in practice just a continuation of it.

As well as the problems inherent in the administration's structures, there are conflicts in the day to day running of the administration between traditional culture and the principles of

---

7. I.M. al'-Awaji, *Bureaucracy and society in Saudi Arabia,* unpubl. Ph.D Thesis, University of Virginia, 1971 p.130-1.

modern bureaucracy. Some of the administration's inefficiency is due to the importation of traditional duties and priorities into a 'rational' administration. In a survey carried out for his thesis, al-Awaji found that civil servants considered that a) their own and their families' interests came before public interest; b) favouritism towards their family in their work was a duty; c) nepotism was seen as a normal way of behaving and was in no way reprehensible; d) bad administration was the chief hindrance to development; and f) they admitted to wasting much time receiving visitors.

The same author identified the following as commonplace problems: the lack of qualified personnel; the frequent irrelevance of a person's training to the post he occupies (he mentions cases of people being given jobs for which they were totally unqualified, such as the geologist who in the early 1960s was appointed director or a hospital); widespread 'corruption' which at the lower levels is due both to an attitude which sees the state as a constant provider and to salaries which cannot keep up with the rise in the cost of living; and a lack of 'public spirit' — the habit of wasting time reading papers and receiving visitors, thus postponing work to a constantly further removed future date.[8]

It is within this administrative confusion that Saudi Arabians at all levels of education are finding employment. As we have discussed the indigenous population is particularly keen to obtain employment in government service rather than in other sectors because of the status attached to it. Low level government positions are therefore filled with young people whose families are not members of the privileged classes; having obtained an educational certificate, these young people have not taken the opportunity of furthering their education as they prefer to start earning, often because their families are dependent on them. They tend to be either unconcerned with the problems facing their country, and in such cases do not care about their level of achievement at work; those who are concerned with the development of the country see the main trends which they are unable to affect in any way. Their alienation may, in the long run, develop into a serious challenge to current development options.

In the higher echelons of the state apparatus there has been a rapid growth in the number of posts, which have been filled mainly by US-trained graduates from comparatively wealthy families. In

---

8. al-Awaji *op cit* p.180-210.

recent years these well qualified men have returned to Saudi Arabia to take up positions of some weight in the various sections of the administration. Mainly born in the 30s and 40s, these people were brought up in Saudi Arabia in the 1940s and 1950s and have had the opportunity of living both in the very restrictive social milieu of their childhood and in the USA, the most advanced capitalist state. They were brought up in a totally undeveloped country and are in a hurry to give advantages to their people. The personal transition such people have gone through was illustrated by the Saudi Arabian Minister of Industry, Ghazi al Qusaibi:

> 'I was born in 1940. When I was born, my mother died of typhoid because there was no doctor in town. Later I almost lost my eyesight because there was no eye doctor in town. I went to primary school without benefit even of candles. I don't believe I saw soap until I was more than 10 years old ... unless you take into account the yearning of our people for a human existence after more than 3,000 years of inhuman existence, you can't understand why we hurry so much.'[9]

The physical restrictions mentioned in this quote must be added to the moral and religious attitudes which pervade every moment of existence. The fundamentally schizophrenic experience of life both in Saudi Arabia and the USA has deeply influenced these men's outlook on the possible directions in their country's social and economic development could take. To some extent, they have internalised both the demands of traditional society of hospitality, assistance to family and tribe, and on the other hand the American standards of efficiency and technologically oriented administration.

At the social level this group suffers from the same externally imposed contraints as the rest of society, namely the Wahhabi values of Saudi Arabian society which they find difficult to reconcile with the liberal and 'dissolute' habits they have acquired in the west. They are the group most likely to put pressure on the regime to liberalise social life, by giving women more rights and abolishing the prohibition laws, as well as relaxing religious compulsion. At the politico-economic level, their concern tends to be to speed up 'progress' and the reconciliation of free-enterprise

---

9. *Newsweek*, 6.6.77.

dependent on capitalism with the traditional monarchic-tribal political structure.

The situation in the late 1970s is that although most of these people are in a relatively privileged situation, they still have little political power. Their only means of influence is through alliance with one of the factions within the royal family, i.e. the one which supports modernisation and the continuation of western influence. Even ministers who are commoners have only very limited power derived from their position and must rely on access to the royal family.

Although the interests of the western educated technocrats are clearly not in conflict with those of the modernist faction in the royal family, there are points on which they are in a greater hurry for changes. These are mainly social issues of liberalisation of personal and public life. The reforms which they want, which cannot at present be conceded because of pressure from the religious authorities, will eventually have to be conceded, when the traditionalist religious faction is weakened.

The Rising Bourgeoisie and the Royal Family

Arising mainly from the traditional Hejazi trading families, this group has expanded its wealth and interests and entered politics. This stratum now includes the small businessmen of the East who flourished thanks to Aramco's support, and those who have established themselves in business through their connections with the ruling family: these are mainly men whose fathers or other relatives held significant positions in the royal household

Initially largely a comprador bourgeoisie, insofar as most of the companies they own acted merely as intermediaries between foreign producers and local consumers, this aspect of their business has remained very important. Many of these companies are the agencies for well-known manufacturers and are importers of anything from watches to missile defence systems. Their rates of profit are such that they need look no further.

The most visible trading companies are those set up by men who owe their rise to favours from the royal family. The most notorious of these is Adnan Khashoggi, son of Abdel Aziz's doctor, who has acted as middleman for all kinds of deals, taking heavy commissions on the way. He has taken large commissions for his dealings and came to the attention of the western public over the outcry over the Lockheed and Northrop bribes to help them sell military aircraft. Thanks to the vast sums he has thus gained (he claims $575 million between 1971 and 1977) he has set up a number of companies, the main one being TRIAD, which he uses to buy up anything from banks in the USA to ranches in the Sudan.

It is clear that the majority of Saudi Arabian businessmen are content to remain in the import/middleman role, but a small number of large and small enterprising companies are now investing in industry.As discussed in Chapter VII, this has taken place mainly in the building sector. For example the AliReza family, which started trading centuries ago, selling frankincense and silk, diversified their interests in the 1950s into the import of Ford and Lincoln cars and Zenith radios, established themselves as agents for numerous companies and have now become one of the biggest building firms in Saudi Arabia by participating in a joint venture with the British firms Laing and Wimpey. One member of the AliReza family is now the Saudi Arabian ambassador to the USA. A similar case is the Juffali company, who are agents for about 20 major western companies, and have set up over 12 joint-venture companies in oil related and building industries. Although doubts may be cast on the permanence and long-term value of the industries that are being established, they certainly are a move forward from the total dependence on imports which the concentration on trade implies.

The dominance of a comprador-type bourgeoisie over an industrial bourgeoisie in Saudi Arabia is not the product of the same factors as in other third world countries. In Saudi Arabia the finance is available for investment in productive industries, what is lacking is the labour power and the raw materials. But within the external limits imposed on independent economic development, it would be possible to set up more industry than has been attempted.

The dominance of trading is due partly to the lack of incentive to invest in other sectors: first the state is able to provide the population with a substantial income without any productive activities and secondly, it is more immediately profitable for a bourgeoisie to import consumer goods than to industrialise and have to compete with the most advanced industrial techniques. This group is also under considerable pressure from the various Western companies who need their agents and partners to ensure maximum exports to Saudi Arabia. It is illegal for western firms to operate in the country without an indigenous agent. Thanks to the growing number of people who have large incomes, the demand for consumer goods is extremely high, and the demand for imports similarly so. It is clearly against the interests of western exporters to see Saudi Arabia develop a self-sufficient industrial and agricultural base, and the advisors, both of the government and of the businessmen encourage the development of a consumer market rather than an industrial base.

Government attempts to encourage private investment in productive industry have had little success, partly because profits are not immediate, or even guaranteed, and partly because some of the projects have been failures, as they have not functioned properly or the products successfully competed with imports. As a result, the state has been obliged to initiate investment itself in the hope that profits will attract private business at a future date.

The rising bourgeoisie on the whole has a vested interest in the continuation of the present social and economic system and is likely to support the continuing trend towards modernisation and westernisation in any political conflict which may arise. Although it feels culturally constricted by the religious restrictions in the country, the extent of change it would support is limited to the field of cultural liberalisation.

The ruling family are not just the people who control the state. Thanks to Abdel Aziz's numerous marriages, many of them to acquire the support of tribal leaders, he left behind over 40 sons and an unspecified number of daughters. The family includes not only these people and their descendants but also those of Abdel Aziz's brothers and other lineages of the Al Saud. Most of the older generation are uneducated, but the younger generation, including the sons of the late King Faisal, have been to university, usually in Britain or the USA. The most prominent member of this younger generation is Prince Saud ibn Faisal, (born 1941) who was appointed Foreign Minister in October 1975, in the footsteps of his father who had been in charge of foreign affairs since 1930.

It is necessary to include in the royal family those linked to it, by kin or marriage, such as the Jiluwi, the Sudairi, and the Al-Shaikh; the latter are the descendants of Abdel Wahhab and keep control of religious ministries and other related matters. In 1977 three of them were in the Cabinet in control of the Ministries of Justice, Higher Education, and Agriculture and Water.

As well as wielding political power, the royal family in its widest sense is the most powerful social group in the country.

Many of its members are involved in business as partners in joint ventures, or sponsors of companies: for example one of the seven Sudairi brothers, Abdel Rahman, looks after the Sudairi family's financial affairs and is known among the foreign community as Prince Moneybags. The family is also the most significant consumer unit as all members, including infants, receive allowances from the privy purse, whose size is undisclosed, but the allowances are known to be substantial (possibly $40,000 a month for top princes) and allow the members to maintain a luxurious standard of living.

The upper classes of Saudi Arabia have attracted far greater public attention in the west than have other developments in that country. This is largely due to the media's attraction to stories involving corruption, sex, gambling and conspicuous consumption. The most notorious features of this have been the quantities of newsprint devoted to Khashoggi and his 'business methods' and to the frequent discovery of members of the Royal Family in gambling houses losing vast sums (Crown Prince Fahd is said to have once lost $6 million in one night), as well as to stories of call girls and royal princes. A further aspect of this is the induced xenophobia in the form of 'the Arabs are buying up London', with headlines on the purchase of stately homes in Britain and Europe, and after-hours shopping trips in Harrods, including purchases in the million dollar range.

Although it is clear that many of these things are happening, they are by no means the most significant aspects of Saudi Arabian society today. When decrying conspicuous consumption and waste, it is worth remembering that only 50 years ago, Saudi Arabia was one of the most destitute parts of the world and that contact with western consumer goods and encouragement to indulge in wasteful expenditure has come largely from westerners trying to build a market in the country. The other aspect of this is that the Saudi Arabians, when they visit the west, witness the extent to which waste prevails there, and the carelessness with which valuable resources, such as oil, are wasted in capitalist society, with no planning for future generations. It is little wonder that they feel entitled to luxuries which, measured in terms of their income, are minor. The real cause for complaint is not that the wealthy Arabs are living it up at the expense of the west, but that while this wealth is being frittered away, there are still people in Saudi Arabia who are poor, have no roof over their head and cannot afford to feed themselves adequately. Further afield there are millions of 'brother Arabs' who live in abject poverty.

The Position of Women

It is notorious that the position of women in Saudi Arabia is abominable. Officially, women are restricted to their homes, not allowed to go out except veiled and accompanied, not allowed to drive cars, or to work except in all-female institutions. It was only in 1960 that girls' schools were first set up but ironically they have been run by the religious authorities, the people most opposed to any form of emancipation of women.

In the nomadic and agricultural communities, women held a place of honour. As well as being in the centre of the household, and responsible for production in it — not a negligible task within a pre-capitalist economy, they participated in agriculture and herding outside the home. They assumed a fair share of responsibility for the survival of the group. Although Islamic customs may have been theoretically in force, women in agriculture and nomadic society had the opportunity of meeting people beyond the narrow confines of the immediate family. Young women met men at wells, and nomadic women were responsible for hospitality towards any passing stranger when the men were away from their tents. Because of the integration of the household and the economy, women were not totally isolated from the world of men. The present process of disintegration of traditional agricultural and nomadic structures is affecting women in a particularly destructive way, as they do not have the options open to men. Although they may be able to obtain some education, they cannot join the labour force and earn money except within the very narrow confines of women only institutions, or in domestic service. Nomads and peasants do not see the point of sending their daughters to school as there is no obvious benefit, such as is apparent for boys' schooling. The attempt to maintain traditional handicrafts in the Social Development Centres is one way of providing women with limited opportunities for entering the productive sector but is of little significance at the national level.

The restrictions imposed on urban women are gradually being eroded by the process of change. For example, since the early 1960s women have been allowed to introduce women's programmes as well as to sing on the radio, both of which were previously banned. Once television was installed, it soon became necessary for women to appear on these programmes, though at first only their hands were to be seen on cookery programmes. No official ruling has formalised women's right to appear on television, but practice is relaxing very slowly to allow people to get used to the idea.

The major factor which is now affecting many women is the fact that the increased wealth in the country means that many women from the wealthier classes are travelling to the west, either on short visits or to study. There they come across a totally different social experience as women: they see women working publicly, behaving freely in the streets and shops, studying in libraries, going to the cinema. This experience is bound to influence their thinking and their aspirations in life, and give them the incentive to demand change in women's position in Saudi Arabia.

There have also been changes in the influence and situation of women according to the political developments in the regime. When Faisal took power in 1962 and felt the need to seem more progressive because of the pressure of Nasserism and the dissatisfaction in the country, it was useful to him to allow a movement for women's emancipation to be vocal and achieve some demands, such as being allowed to work in the new Ministry of Labour and Social Affairs. There was also at that time some discussion in the press of such matters as arranged marriages, polygamy, early marriage and the lack of educational and work opportunities for women. In 1964, women were given scholarships to study abroad and the opportunity to work in women's charitable organisations or even in government institutions.

But the advances were halted dramatically in 1970 when the regime felt strong and secure against external threat. As Shaker notes:

> 'Saudi women also had their observations of the "socio-political" situation at that time. The scope of their activities was greatly reduced as of 1970. As was demonstrated .... during 1962-4 the Saudi woman emerged as a viable element in the country's surge towards modernization. This was evident in her new role in the Ministry of Labor and Social Affairs, the radio and the press in addition to her traditional role as a teacher in the area of elementary education. The Saudi woman during that period also took the initiative in establishing a number of "Women's Organisations" across the nation. But most significant was her leading role in calling

for and giving support to the establishment of a women's college as part of the University of Jiddah — that was to be a private university. During 1970 most of the women's activities in these areas were liquidated. Women's employment and presence in the Ministry of Labor and Social Affairs were reduced to a nominal status. There were also doubts as to the survival of the Women's College in Jiddah.'[10]

According to this author's informants, this regression had a number of causes, mainly the opposition of the religious movement, whose power at that time appeared to be growing. The other explanation was that King Faisal was unwilling to support women's demands against such opposition and had not attempted to defend the women against the religious authorities. The exclusion of women from the male dominated administration is made particularly obvious by the fact that the modern ministries and other official buildings have no provisions for women's toilet facilities.

In recent years the conflict has intensified between the need for a larger labour force, and consequently the incentive to allow women out of the home and into the economy on the one hand, and on the other the religious custom of keeping women secluded from men and within the realm of the home. Development policies and the fear of being swamped by immigrants are likely to force the regime to adopt a more liberal position on the employment of women. The problem is that both the employment of women and the importation of foreign (particularly non-Muslim) labour force go against the principles of the religious authorities. For them the choice is one between two evils. In response to these pressures women are currently allowed to work — but in women-only institutions: they are found at all levels in women's hospitals and schools up to university level. In the universities which educate both men and women, the students are totally segregated: women who have to be taught by male teachers are taught through close circuit television. Libraries have separate opening hours for male and female students.

Another problem women face is the country's underpopulation. The regime's determination to increase the birthrate has led it to

---

10. F.A. Shaker, *Modernization of the developing nations: the case of Saudi Arabia* unpubl. Ph.D Purdue University, 1972, p.315-6.

ban all forms of contraception. This policy tends to keep women in the homes in order to enlarge the next generation, and although these laws may be effective with the poorer, uneducated women, the upper class women are still able to import contraceptives when they visit the west.

Finally the tendencies reducing the family to nuclear units are further isolating women since their movements become limited to houses where they live alone with their husbands and children and lose contact with the extended family. This is in some way compensated for by two factors: first the increased distribution of telephones gives women the opportunity of communicating with each other; and the development of radio, television and magazines which, although they are largely aimed at increasing the consumer market, do give women some distraction from the drabness and boredom of their newly isolated lives. For some highly-educated couples, there is a greater degree of cooperation and communication between themselves, which helps to replace the close contacts women used to have with the extended family. In the long run, this new type of companionship within the couple may assist the development of close nuclear families on the western model, but in fact marriages do not tend to last long, despite the reduction of polygamy. Although Saudi Arabians now have adopted an architecture which encourages the nuclear unit, they have little inclination towards it.

The Process of Urbanisation

As we have seen, the present trend in Saudi Arabian society is for a rapid depopulation of the countryside be it settled or nomadic, in favour of an equally rapid expansion of the cities. The main centres of expansion are the capitals Riyadh and Ta'if, the main commercial town Jeddah, the pilgrimage centres Mecca and Medina, and the oil conurbation Dammam-Al-Khobar-Abqaiq. Their combined population was estimated at 2.3 million in 1975 and is expected to rise by between 30 and 80 per cent in the next fifteen years. Although all figures are to be taken with the caution necessary in a country where census figures are kept secret, there is no doubt about the trend.

This rapid urbanisation has three main aspects: first, most of the people who arrive looking for work are not familiar with the urban way of life and the means to deal with its problems. They arriv destitute and, as these cities are the most expensive places in or proved their advantages over the centuries, modern building is run on Western principles, by Western architects and building firms who are ignorant of local conditions and appear more concerned with their rate of profit on each bag of imported cement and each air conditioning unit than on building sound and livable housing. All over the place concrete modern buildings are going up which do not fit into the region aesthetically, but, far more seriously, are totally unsuited to the climate. Traditional housing had developed the methods to ensure maximum coolness in the extremely hot summers, through a system of air vents, cool retaining bricks, latticed windows, etc.... These are being unfeelingly demolished to make way for concrete skyscrapers, which are meant to represent progress!

This destruction of the old cities, particularly of Riyadh and the smaller towns which are in expansion, is the most obvious physical indication of the way the country is turning its back on the past.

From a pre-capitalist society half a century ago Saudi Arabia is now rapidly developing into a class society within the limits imposed by its economy based on the export of oil. Its ruling class is composed primarily of the royal family, and is slowly being penetrated by the local bourgeoisie involved in trade and to a minimal extent in industry. The indigenous population is increasingly composed of a substantial group of technocrats and office employees, while it includes almost no working class whatsoever. The vast majority of people engaged in manual labour, in construction and minimally in industry, are migrant workers. The only really poor section of the native community are the peasants and nomads who have been unable to capitalise their activities and whose economic base is in decline. This generation is being driven to the cities as unskilled labour, but in the long run, it seems that most Saudi Arabians will get an education and therefore find openings in the administration or for those who acquire technical skills as technicians in the future capital intensive industries. Whether Saudi Arabia's dependence on a migrant labour force is to be a permanent feature will depend on the type of non-oil industrial development which emerges in the coming decades: so long as local industry is non-existent or highly capital intensive and the regime trains enough educationalists, doctors and bureaucrats, as well as maintenance technicians, there will be little need for migrants. On the other hand while expansion of the infrastructure and labour intensive activities continue, then migrant workers will continue to be necessary. The position of migrant labourers in society is unlikely to improve, as apart from Yemenis, the regime has no reason to encourage any of them to settle on a permanent basis.

Social liberalisation is bound to come about eventually, although it will be associated with the defeat of traditional values and the victory of western corruption.

CONCLUSION

As we have seen, radical changes have taken place in Saudi Arabia in the last twenty years. Oil revenues have allowed the regime to assert its control over the country as they have enabled it to consolidate its political hegemony by paying subsidies to the tribes and by buying off any potential oppostion. For the same reason the regime is in control of economic development and, less directly, of social transformation.

One reason why Saudi Arabia's situation is unique is that the country's wealth has not come as a result of the development of the country's internal productive forces, but from the sale abroad of a raw material whose exploitation requires few workers, only about 20,000. Although oil is a Saudi Arabian resource, its exploitation is effected through external agents. The income generated from oil is paid directly to the regime from abroad, it forms a direct support for the political *status-quo*. Oil income has not been *the product* of internal economic development, on the contrary it is *its cause*. This reversal of standard principles of development has deep consequences for Saudi Arabia socially, politically and economically.

Economic activity is rapidly departing from the traditional resources of herding and agriculture. The only traditional sector which is in expansion, and incomparably so, is trading, particularly imports. Oil income has made it possible for the country to live off imports of everything from wheat to private aeroplanes. The expansion of the trading sector is taking place with the active encouragement of the regime which has taken steps to create a market for the importers.

This is happening partly through the creation of consumers: the state employs people at increasingly large salaries to work in the administration and services, initiating a consumer society. The other ways in which the regime has supported the development of a merchant class have been its own extravagant consumption and the tacitly acknowledged 'commission' system of agencies which has allowed a number of fortunes to be made with surprising speed. Even in trade, the apparently most independent sector, the regime's influence has been crucial in assisting expansion.

The origins of the regime's full control over the country also explain why economic development, in the traditional sense, is hardly taking place. Formally the regime encourages the development of private enterprise in industry and agriculture. Up till now there has been little investment in those sectors. The country continues to be increasingly dependent on imports and the development of indigenous industry has been insignificant. The only field in which indigenous capital seems to have been involved successfully has been building, which has benefitted from the government's massive infrastructure expenditure and provides opportunities for rapid profits with a minimal investment. The current Second Plan involves primarily building projects, be they of roads or hospitals, and the only industrial projects of any seriousness also included are oil related. These plans are being postponed further into the future as technical hitches accumulate. Their future is uncertain but the petrochemical projects may provide the basis for an industrial future.

Non-oil industries only play a marginal role in the economy, as imports have been found to be more immediately profitable to the merchant class. The long-term future of the country's economy is still open as the present trends are not yet really set and a change of course is still possible. But the prospect for the creation of an autonomous economy are limited by the shortage of labour and raw materials and therefore the need to invest in capital intensive industries. This, in turn, means dependence on technologically advanced states such as the USA and western European countries, to supply the required machinery.

This apparent dependence of Saudi Arabia on capitalism, both for capital and consumer goods is often regarded as the reason for its equally apparent subservience to US policies in international affairs, particularly on such issues as Palestine and the Gulf. This interpretation is a misunderstanding of Saudi Arabian policies. Although, in the past, the US certainly could put more pressure on Saudi Arabia than vice versa, following the oil price rises of 1974 Saudi arabia has accumulated surplus hard currency revenues to such an extent that, within a few years, it might even be able to keep its population comfortably in luxuries without producing a single drop of oil, simply from the income of its foreign investment of surplus revenues. These investments are almost totally in the west.

This situation clearly gives the Saudi Arabian regime a vested interest in the survival of capitalism thoughout the world and explains its willingness to make what it sees as minor compromises in the short run to ensure the long-term continuation of a monarchy in Saudi Arabia. In this sense, the country can be said to be on the way to becoming a rentier state insofar as the regime could live off overseas investment and the population may no longer have to rely on any internal production at all. It also explains the regime's continued violent opposition to communism, the only potential real challenge to the present course. As we have seen there are other, religious, reasons for the regime's opposition to communism, which in Saudi Arabian terms is not defined scientifically but merely abhorred because for its atheism. This pathological hostility to 'communism' — a term used to cover all evils — explains the alliance and convergence of interests with the USA in international politics, both in the Third World and in the West.

The internal political situation is a different story. To ensure its continued control, the royal family has insisted on the enforcement of strict Wahhabi practice, as it feared deposition, as has happened to other ruling families in the Middle East. Based on a distortion of Wahhabi ideology the regime has consistently claimed to uphold its values and has used it as a means for total ideological control. Since Wahhabism is a call for the return to the fundamentals of Islam and an unquestioning acceptance of social rules laid down in the Koran, enforcement of this ideology naturally means that neither political nor religious developments consistent with the technological advances of society can be countenanced. Such a position forbids any intellectual or political life and repression within Saudi Arabia both by brute force and more subtly at the educational and cultural levels. The total exclusion of foreign ideas, Arab or otherwise, has meant political and intellectual sterility.

The regime also has other weapons to combat opposition, particularly its ability to buy off any potential hostility long before it has time to develop into anything like a serious threat. Finally the process of social transformation which has taken place has not yet given birth to developed class forces with antagonistic interests within the society. Lacking a developed working class there is little likelihood of a serious opposition to the present autocratic regime. Migrant labourers are not significant in the argument as they are particularly vulnerable and have little interest in organising to overthrow the regime. The educated groups have access to positions and high incomes.

We thus have the strange phenomenon of a state which is an anachronism in political terms in the Twentieth Century, while at the same time it is a financial world power and has introduced massive technological modernisation in the country. It is a breeder of rampant materialism while it bears the cloak of asceticism. It is a place where freedom of speech, thought and action do not exist, where the gap between wealth and poverty is gigantic. Yet the regime subscribes to fundamentalist Islamic ideas by which every man is equal. It is a ruling force in the Arab world because of its wealth; yet its people are forcibly cut off from the cultural and social developments which have taken place in the Middle East in the last twenty years.

Social transformation is the most significant long-term effect of this economy based on oil income. Financial wealth superimposed on a totally undeveloped society has created a life based on the material comforts of the west, or an imitation of them, imported as a result of the 'modernising' influence of the USA in particular. The sudden material transformation of a way of life which had remained at a pre-industrial stage for centuries, cannot fail to have a deep impact on the society.

Wahhabism has been retained as the ideology of a regime which is introducing this massive and immediate change in its people's way of life.

Within twenty years or less people have moved from desert tents with bushes as lavatories to villas with gold plated bath taps. The ideological confusion which arises from this contradiction is enormous. Western producers and local importers encourage the development of materialism and even King Faisal, the religious ascetic, favoured material modernisation. But the overflow of wealth relative to a decade ago has meant that the consumer society is developing on an unprecedented scale and the people of the desert have become gripped by a materialism verging on the philistinism of US society. The rule of the commodity has brought the final destruction of the desert ethos of thrift and self-sufficiency.

This consumerism has developed at a superficial level insofar as it involved neither the development of technology nor of a culutre, depending merely on traders expanding their range of agencies on the one hand, and the continued control of the Saud family on the other. As we have seen above the Al Saud have good reason to retain Wahhabism as the base of their political control as it prevents the development of independent thinking, dangerous to their rule, and has given them the opportunity to develop their ascendancy in the Muslim world. Here Wahhabism has been the product, and financial power the means.

One reason why the regime is so opposed to external intellectual influences is that the continued equilibrium between the reality of materialism caused by rapid westernisation and the fiction of Wahhabism which has lost its real roots with the destruction of the age-old desert culture can only be maintained by an intellectual petrification. It is here that one can see the validity of the positions maintained by the traditionalist faction of the ruling family and its supporters. Their desire to slow down development is in a way the only thing which could retain the traditional culture as such a slowing down could give the people the opportunity of creating a Saudi Arabian culture consistent with the developments in the country. The present headlong rush into consumerism and US orientated development policies as well as the increase in western building techniques etc, are causing cultural destruction without offering a replacement. The culture of the desert gives way to a cultural desert.

Such a situation cannot last and the immediate prospects for Saudi Arabia are bleak. The ruling family are bound to support the shell of Wahhabism as it is necessary for their control of the state, while pursuing further 'modernisation' policies which may give the people some material improvement but lock them in a contradiction between the ideology they are brought up with and the extravagant materialist daily life they are expected to aspire to. The forced intellectual sterility can only temporarily prevent the emergence of a new culture.

FURTHER READING

The following list includes only basic background books and a few interesting analyses of change in Saudi Arabian society. It is neither meant to be a comprehensive bibliography nor to include all the works used in writing this book, but only to give the general reader the opportunity of looking at some interesting recent material. A full bibliography up to 1970 can be found in the *Area Handbook*.

I.M. Al-Awaji *Bureaucracy and Society in Saudi Arabia*, unpublished Ph.D Thesis, University of Virginia 1971.
Aramco Handbook, Oil and the Middle East, Dhahran 1968.
Area Handbook for Saudi Arabia, American University, Washington DC 1971.
D.P. Cole *Nomads of the nomads,* Chicago, 1975.
M. Katakura, *Bedouin Village,* Tokyo 1977.
R. Knauerhase, *The Saudi Arabian Economy,* New York, 1975.
E.A. Nakhleh, *The United States and Saudi Arabia,* Washington DC 1975.
F.A. Shaker, *Modernization of the developing nations: the case of Saudi Arabia,* unpublished Ph.D Thesis, Purdue University, 1972.
D.A. Wells *Saudi Arabian Development Strategy,* Washington DC 1976.

INDEX

Abdel Aziz (King): 14 et seq, 134, 169
Abdel Mohsenibn Abdel Aziz (Prince): 91
Abdel Mohsenben Jiluwi: 102
Abdel Rahman ibn Abdel Aziz (Prince): 101, 201
Abdullah (Prince): 73
Abdullah Aysi: 102
Abu Dhabi: 112, 128
ADC (Arabian Drilling Co): 54
AGIP: 54
Agriculture: 3, 147, 156, 183
— table of crops: 183
Agricultural Bank: 157, 184
Ahmad Tawil: 103
Air Force — see Saudi Royal Air Force
al-by Said: 113
AliReza family: 203
Ali Zein Abdine: 102
Anglo-Persian Oil Co — see BP
Anwar Ali: 192
Arab Destiny: 103
Arab Federation: 113
Arab League: 41, 117 — peace keeping force: 123
Arab Nationalism: 62, 76
Arab Oil Congress (1st and 2nd): 41
Aramco (Arabian American Oil Co): 37, 60, 138, 196 — profit sharing agreement: 37, 46 et seq
— strikes: 95 — & the October War: 120 — gas gathering: 159
ARGAS (Arabian Geophysical & Surveying Co): 55
Armed Forces — see Saudi Armed Forces, National Guard
Artawiyah: 23, 26
AUXIRAP (Societe Auxiliare de la Regie Autonome des Petroles): 42, 54
Asir: 29, 82, 83, 185

Assad (President of Syria): 70, 118, 119
al-Awaji: 199
Asian labour: 194
asphalt: 142

Ba'ath Party: 105
Badribn Abdel Aziz (Prince): 91
Bahrain: 72, 128
Bangladesh: 130
BAPCO (Bahrain Petroleum Co): 32, 35
Bechtel Corp: 133, 159
Bhutto (President of Pakistan): 129
ibn Bijad ibn Humayd: 23
bedu (bedouin): 40, 174 et seq
Black September: 120
borders: 18
Britain: 14 et seq, 134 — 1915 Treaty: 20 —'s Royal Air Force: 25
— subsidy: 30 — footballers: 196
BP (British Petroleum): 32
building (modern): 215
Buraimi Oasis: 76, 112
bureaucracy: 197

Caltex: 35 et seq
censorship: 35 et seq
Clayton, Sir Gilbert: 19
commerce: 202
Committees to promote Good and suppress Evil — see Public Morality Committees
communist countries: 134
Communist Party: 104
Community Development Centres: 84
Constitution: 91 et seq
consultation: 92
Council of Ministers: 59
Council of Senior Princes: 93, 114

Coups, attempted: 102 et seq
Cox, Percy: 17

Damman: 87, 145, 150
desertification: 169
Daimler-Benz: 162
demonstrations: 101
Development Plans: 82, 137 — tables 140, 153 — Second: 142, 148
Dhahran: 35, 37, 91, 97, 101, 112
Dhofar: 127
Dickson: 25

Education: 79, 163 — women's: 80 — ratio of boys & girls: 79
Egypt: 64, 67, 73, 91, 100, 113, 130
Eisenhower Doctrine: 113
electricity service: 145
embargo — see Oil
employment: 144
'Enemies of God': 91
Eritrea: 73, 126

Faisal ibn Abdel Aziz (King): 62, 64, 92 — ten point programme: 64 assassination: 68 — foreign policy: 114 — women's emancipation: 209
Fahd ibn Abdel Aziz (Prince): 70, 72, 73, 102, 127
Faisal ibn Musa'id ibn Abdel Aziz: 69
farms, experimental: 186
Fatima (wadi): 186
Fertile Crescent Plan: 123
Fluor Corp: 159
Ford Foundation: 193
foreign aid: 130
foreign policy: 110, 124 et seq
Free Princes: 60, 90 et seq
Front for the Liberation of Arabia: 63

General Motors: 162

Ghamidi, Ali (Director of Investigation): 98
GhutGhut: 23
al Gosaibi: 35
Grievance Board: 76
Gulf coast: 6 — Security Pact: 127

Habib, John: 20
Hadhramaut: 126
hajj: 20, 30
al Hamdan, Abdullah Sulaiman (first Minister of Finance): 29, 59, 186
al Hamdi, Ibrahim (President of YAR qv): 192
Haradh — King Faisal Model Settlement: 179
Harvard University: 151
Hashemite family — see Hussein
health: 81, 146, 163
Hejaz: 5, 8, 57, 83, 182
al Hasa: 3, 29, 83, 187
hijra (pl *hujar*): 23
Hofuf: 54
Holmes: 30, 33
hospitals: 82, 146
Hussein ibn Ali (Sharif Of Mecca): 15, 16, 110

Ickes, Harold L: 36
Ikhwan: 16, 18, 20, 21 — settlements: 23
Immam of Yemen: 18, 64
IMF (International Monetary Fund): 62
incomes: 169
industrial development: 138, 159 et seq
International Islamic Conference: 115
IPC (Iraq Petroleum Co): 32, 33
Iran: 118, 126, 160 — Shah of: 115
Iraq: 110, 115
Islam: 2, 7 et seq, 124

Israel — elections 123, — see Palestine

Japan: 134
al Jazira al Jadida: 105
Jeddah — hospital: 82 — port 58 — palace: 59 — steel mill: 138 treaty of: 19 — water supply 156
Jeddah Oil Refinery Co: 55
ben Jiluwi, Abdel Mohsen (Director of Investigation): 100
Jizan: 101, 185
Jordan: 110, 119
Jubail: 17, 145, 159
Juffali Co: 205
ibn Juluwi: 16, 26, 58, 204
14 July Revolution: 113
June War: 101, 116 — see Palestine
Jungers, Carl: 120
justice: 74

Kissinger, H: 70
Khaddam, Abdel Halim (Syrian Foreign Minister): 119
Khaled ibn Abdel Aziz (King): 60 et seq, 92 — foreign policy: 122 et seq
al Kharj agricultural project: 181, 185
Khartoum summit conference: 116 — embassy seized: 120
Khassogi, Adnan: 132, 205
Korea, South: 194
Kutum — Talah (Asir): 159

Labour: 165, 188 — unrest: 188 — Regulations: 190 — Commissions: 192 — migrant: 193
Laing (Contractors): 203
Lawrence, TE: 19
Lebanon Civil War: 73, 117
Little, Arthur D (consultancy): 151
Likud: 123
Lockheed: 202

Longrigg, SH: 33

Mahmal: 26
Majid (Prince): 73
majlis: 58, 174
Mauretania: 130
McGhee, George: 38
Mecca: 7, 18, 82 — water supply: 156
media: 84
Medina: 82
Ministries: 73, 75, 85
Mitsubishi: 160
Mobil (Socony) Oil Co: 32, 37
Morocco: 130
Mossadegh: 49
Movement of Arab Nationalists: 94, 102
Muhammad (the Prophet): 7 et seq
Muhammad Abdel Wahhab: 10
Muhammad Ali (Ruler of Egypt): 12
Muhammad (Prince): 73
al Murrah: 179
Muscat: 13, 113
Muslim Autonomy: 130
Mutib (Prince): 73

Najd: 3 et seq, 15, 17, 57, 83, 182
Najran: 18. 100
Nasiriyah palace: 59
Nasser, Gamal Abdel: 51, 81, 112, 115 et seq — assassination attempt: 61, 113
Nasserism: 60, 108
National Guard: 70, 78, 93, 132, 176
National Liberation Front: 97, 108
National Reform Front: 104, 108
Natomas Internal Corp: 54
Nawaf ibn Abdel Aziz (Prince): 91
Nayef (Prince): 73
Nazer, Sheikh Hisham: 72, 92
newspapers: 29
al Nidal: 105
nomads — see *bedu*

Northrop: 198
North-South Dialogue: 52

Occidental Petroleum: 56
October War: 47, 50, 120
Oil: 32 et seq — tables: 44, 45
— embargo: 51, 119 — Congress (Arab): 41
Oman: 13, 112, 127
Onassis: 39
OPEC (Organisation of Petroleum Exporting Countries): 40 et seq, 149
opposition: 90-93
Ordinance — for the Protection & Encouragement of National Industries: 139

Pakistan: 73, 129
palaces: 59
Palestine/Israel: 69, 85, 111, 135 see October War, June War
palestinians: 94, 193
PDRY see Yemen, South
Petrolube (Petromin Lubricating Oil Co): 55, 143
Petromin (Saudi Nat. Oil Co/General & Mineral Org): 42
Petroship (Petromin Tankers & Mineral Shipping Co.): 55
peasant farming: 183
PFLO (People's Front for the Liberation of Oman): 125
Pharaon, Ghaith: 162
Philby, H St J: 19, 25
Phillipines: 130
Phillips: 54
pilgrimage see *hajj*
Planning (Central Organisation): 199
PLO (Palestine Liberation Organisation: 117 et seq
Popular Struggle Front/Democratic Party: 105

population: 108
Press Law & control: 85
profit sharing agreement: 158 see Aramco
Provincial Regulation Statute: 199
Public Morality Committees: 28, 60, 75, 85

Qabus (Sultan, of Oman): 127
Qasim (President of Iraq): 115
qat: 3
Qatif: 182, — treaties of: 17, 32
Quraish tribe: 7
al Qusaibi, Gazi: 201

Radio: 29, 86
railway: 29
al Rashid: 13, 17, 90
reactionary politics: 116
REDEC: 162
Red Sea States: 125
Red Line Agreement: 32 et seq
Riyadh: 58, 91, 101 — '76 Conference: 117
road building: 145
Royal Guard: 78
RTZ (Rio Tinto Zinc): 158
Rub' al Khali: 54

Sa'ad ibn Faisal (Prince): 73
SABIC (Saudi Arabian Basic Industries Corp): 56
Sadat (President of Egypt): 118
SAFCO (Saudi Arabian Fertiliser Co): 55, 160
SAMA (Saudi Arabian Monetary Agency): 59, 62
SARCO (Saudi Arabian Refining Co): 55
Sana'a see Yemen

Index 223

Sarraj, Col, Abdel Hamid: 113
Saqqaf, Omar: 119
Saudi Armed Forces: 75, 132
Saudi Arabian Tankers Co: 39
Saudi Arabian Communist Party: 104
al Saud family: 11
Saud ibn Abdel Aziz (King): 59 et seq, 90, 96 — foreign policy: 112
Saud ibn Faisal (Prince): 73, 204
Saud ben Jiluwi (Governor eastern Province): 98
Saudi Royal Air Force: 77, 100, 102
Saut al Tali'a: 105
Sbila: 27
sejm alabeed (slaves' prison): 98
September Revolution: 97
settlement of bedouin: 178
Shah see Iran
Shaker: 201
Shakespear, W: 17
Sharia law (Koranic law): 75
al Shaykh (successors of Abdel Wahhab): 75, 204
SIDF (Saudi Industrial Development Fund): 161
Sinclair Arabian Oil Co: 54
Sirhan (wadi): 18, 178
SOCAL (Standard Oil of California): 33, 135
Socony see Mobil
Somalia: 126
Social Development Centres: 206
Social Security Law: 83, 89
Society for the Liberation of the Holy Soil: 101
Sons of the Arabian Peninsula: 97
Standard Oil (Esso, New Jersey): 32
Stanford Res Inst: 151
State Security Law (1961): 97
strikes: 90, 95
Sudairi brothers: 70, 73, 93, 204
Sudairi tribe: 29
Sudan: 73, 130 — see Khartoum
Suez: 112, 160
Sultan (Prince): 73

Ta'if: 18, 25
Talal ibn Abdel Aziz (Prince): 63, 91
Tanura refinery: 142
Tariqi, Abdullah: 40, 63, 75
Third World: 129
Tihana plain: 3, 182
telecommunications: 145
television: 86
Ten Point Programme: 64
Texaco: 32, 33
trade balance: 134
Trucial Oman Levies: 112
Tunisia: 130
Turkey: 14 et seq
Turaif: 158

UAE (United Arab Emirates): 52 et seq, 128, 160
UAR (United Arab Republic): 113 see Egypt
USA — lendlease: 36 — treasury: 38 — legation: 37 — air base: 112 — special relationship: 131 et seq
USSR: 51, 67, 134 — expulsion from Egypt: 118
'ulema: 74, 85
UN (United Nations): 89
Universal Declaration of Human Rights: 89
urbanisation: 209
US Corps of Engineers: 86, 132
US-Saudi Joint Commission on Economic Cooperation: 157
'Utayba: 23

Vinnell Corp: 78, 131, 195
Voice of the Arabs: 60, 85, 100

wahhabism: 10 et seq, 74, 148, 201
water: 155
West Bank State (Palestine): 123
White Guard: 78

Wimpey: 203
women: 206
Workers Committee: 97, 103
World Bank (Int Bank of Reconstruction & Development): 62
World Wars: 16 et seq

YAR (Yemen Arab Republic, capital Sana'a): 67, 97, 100, 114, 125 see Immam of Yemen

Yamani, Sheikh Ahmad Zaki (Minister of Petroleum): 75, 119
Yemen, South (or PDRY, Peoples Democratic Republic of Yemen): 72, 117, 124 et seq
Yemen Civil War: 76, 114
yemeni labour: 191 et seq
Yenbo: 145, 160

zakkat (religious tax): 11, 22
zionism: 113 et seq, 135

A House built on Sand—a political economy of Saudi Arabia
by Helen Lackner

A House built on Sand is the first critical analysis of a society which is playing an increasing part in the world economy. Saudi Arabia has entered the first rank of world financial powers and now exerts enormous influence in international affairs, while at the same time its unique political structure and ideology is derived from its recent history of nomadism, tribal warfare and religious fundamentalism.

This book provides the basis for a better understanding of Saudi Arabia's internal and external situation and corrects some of the many misconceptions which prevail. It charts the rise of the house of al-Saud, its diplomatic and dynastic alliances, and the roles played by oil, religion and the USA in its continued existence. The activities of the political opposition, the role of women, the large migrant population and the vast economic development programme are all discussed.

Helen Lackner studied anthropology at the School of Oriental & African Studies, London University. She contributed to *Anthropology and the Colonial Encounter*, edited by Talal Asad (Ithaca Press 1973). She also translated articles by a number of leading French anthropologists for *Relations of Production*, edited by David Seddon (Frank Cass 1978)

paperback £3.50

Ithaca Press 13 Southwark Street London SE1